Supplement to
"Basic Research in Cardiology", Vol. 80 (1985)

Editors: R. Jacob (Tübingen) and W. Schaper (Bad Nauheim)

Adult heart muscle cells

Isolation, properties and applications

Edited by H. M. Piper and P. G. Spieckermann

With contributions by J. C. Altona, I. A. Bailey, M. Banfalvi,
M. Bechem, Z. Bendukidze, A. Beresewicz, B. Bertier-Savalle,
C. H. F. Bloys van Treslong, J. V. Bonventre, T. K. Borg,
M. Borgers, G. Brum, E. Carmeliet, J. Y. Cheung, J. Eckel,
S. A. E. Finch, W. Graf, D. R. Harris, J. F. Hütter, G. Isenberg,
S. L. Jacobson, H. Kammermeier, U. Klöckner, R. Koch,
B. Koidl, W. Kübler, A. Leaf, E. Lundgren, B. Maisch,
F. Marotte, D. Mascher, A. A. Noronha-Dutra, W. Osterrieder,
F. Pieper, H. M. Piper, L. Pott, T. Powell, I. Probst, L. Rappaport,
B. Rauch, H. Reinauer, J. L. Samuel, R. Spahr,
P. G. Spieckermann, K. Schwartz, P. Schwartz, T. A. Schwarzfeld,
E. M. Steen, A. Stier, L. Terracio, F. Thoné, H. A. Tritthart,
G. Trube, V. von Tscharner, F. Valenzuela, G. Van der Heyden,
A. Van der Laarse, J. Veerecke, L. Ver Donck, B. Wein, N. Woolf,
G. Zernig

Springer-Verlag Berlin Heidelberg GmbH

ISBN 978-3-662-11043-0   ISBN 978-3-662-11041-6 (eBook)
DOI 10.1007/978-3-662-11041-6

© 1984 by Springer-Verlag Berlin Heidelberg
Originally published by Dr. Dietrich Steinkopff Verlag GmbH & Co. KG. Darmstadt in 1984
Softcover reprint of the hardcover 1st edition 1984

Foreword

With this volume results become generally available of a recent conference that was the first to discuss properties of isolated heart muscle cells from the adult myocardium as a new cardiological research tool. The conference was held as an International Erwin Riesch Symposium under the auspices of the Deutsche Gesellschaft für Herz- und Kreislaufforschung at Spangenberg Castle, March 22–25, 1984. It was made possible through funding by the Erwin Riesch Foundation and the Bayer AG which deserve our gratitude for their generosity.

It may not be fully reflected in this collection of papers that the atmosphere of this meeting was unique. The reason for this resided apparently in the still rudimentary experiences with the adult cardiocyte model that is now widely acknowledged to be very promising. Thus, quite basic interests brought together experts from different biological disciplines who in respect to this new model were all still "beginners". It was noteworthy that such common interests exerted a unifying effect upon otherwise rather separated fields of science.

The main task of the years to come will certainly consist in improving isolation and culturing conditions for adult myocytes in order to fully establish the isolated cell model in heart research. Very basic investigations are still to be done to achieve this goal. For this, cooperation and information exchange is urgently needed among the yet limited number of groups working actively with isolated adult myocytes. The broad perspectives of a further development of this model are elucidated not only by the topics of the articles in this volume, but also by the summary of the informal discussions at the conference.

We hope that these proceedings may contribute to the progress in research with cellular models of the myocardium.

H. M. Piper
P. G. Spieckermann

Contents

Isolated adult heart cells.
The development of a new experimental model

T. Powell

Department of Medical Physics and Institute of Nuclear Medicine, The Middlesex Hospital Medical School, London (U.K.)

Summary

The isolation and characterization of calcium-tolerant cells from adult myocardial tissue has proved a major new field of heart research. The wide range of problems which can be examined using single cells in suspension offers the possibility of investigating many important mechanisms underlying diseased states in the myocardium.

Key words: isolated myocytes, calcium tolerant, myocardial mechanisms

1 Introduction

The history of the development of a suitable preparation of adult ventricular cells isolated from rat heart is instructive in terms of both the philosophy underlying the preparative techniques employed and the criteria which have been established to determine myocyte viability. Kono (10) demonstrated in 1969 that organ perfusions with trypsin and α-chymotrypsin followed by purified collagenases yielded a suspension of isolated rat myocytes. In the following year, Berry, Friend and Scheuer (3) described a method using heart perfusion with crude collagenase and hyaluronidase, and Vahouny, Wei, Starkweather and Davis (20) reported that incubation of tissue fragments with a mixture of trypsin and crude collagenase was also effective in producing single cells in suspension. These latter two groups noted that the isolated cells went into hypercontracture on the addition of physiological levels of extracellular calcium chloride, which prompted experiments in the author's own laboratory to demonstrate that a short *in vitro* perfusion with crude collagenase alone was sufficient to produce adequate yields of myocytes (9).

From this simple beginning, more than a decade ago, has evolved an ever increasing literature on isolated heart cells, and the author and his colleagues in London can claim with justification to have pioneered the isolation and characterization of ventricular myocytes from the adult rat heart. During the four years following the initial abstract, detailed experiments were carried out to achieve routine preparations of rat cells, and methods for myocyte purification were also examined. This work culminated in the paper by Powell and Twist (16) which has become a standard reference in the field for several important reasons. First, it specifically highlighted three important advances: that the technique was rapid and reproducible, that the isolated cells exhibited respiratory control, and that the cells displayed a tolerance to calcium. The term "calcium-tolerant" (which the author can claim responsibility for coining) is now accepted as one major criterion to be satisfied if myocytes are to be considered suitable for experimental studies. Secondly, the paper detailed the cell yields to be expected routinely using this method, and reported on the proportion of the preparation ($\sim 70\%$) comprising undamaged cells which are rod-shaped.

Thirdly, the authors considered briefly some of the factors which might have contributed to the production of calcium-tolerant cells, a theme which has been examined in great detail since then (4, 5, 11, 12).

It was not coincidental that in the same volume of the journal containing Powell and Twist's first paper, a second report (17) described the stimulation by isoprenaline of cyclic adenosine 3', 5'-monophosphate in the isolated cells. This second paper was viewed by the authors as a companion manuscript to the first, illustrating that the preparation was of sufficient yield to allow measurement of a biochemical signal triggered by a pharmacological intervention, under conditions where the spatial organisation of the receptor-enzyme complex resembles that found in the intact tissue, without the presence of an extensive extracellular space. In addition, the results obtained using calcium-tolerant cells were contrasted with previously reported data using myocytes sensitive to calcium, to emphasize the importance of studying cellular function under conditions where the extracellular concentration of this divalent cation is within physiological limits.

2 Cell isolation

There is no doubt that the importance of individual myocytes isolated from the adult heart as a model system for investigating the cellular mechanisms underlying cardiac function is now universally recognized. However, one major problem is the dichotomy inherent in the use of any new preparation for definitive experimental studies. On the one hand, the main reason for obtaining individual cells is to investigate myocardial mechanisms which could not be examined adequately using other available preparations, but at the same time it is necessary to establish that, within the context of the mechanism under study, cell function has not been compromised by the preparative techniques. The plethora of dissociation methods currently available, even only for rat heart, claiming "normal" ultrastructure and function for the isolated myocytes merely exacerbates the problem.

One notable feature in the history of studies involving adult mammalian cardiac myocytes has been that "improved" isolation procedures have tended to be more and more complex compared to their predecessors. It is relatively rare for the rationale behind the introduction of modifications to the basic techniques to be either clarified or justified by control experiments. Of more concern is the tendency of many authors not to report the success rates of their isolation procedures, but to infer by omission that their particular method is quite straightforward and reproducible. This problem will become more serious as numerous cell types are isolated from a wide variety of animal hearts. While it is clear that preparation of rat or guinea-pig ventricular cells in high yield is now considered routine in London, for example, future research will answer many important questions concerning whether or not myocyte preparation will become a routine laboratory technique, or remain for some time a state-of-the art procedure, where specialized knowledge and control of the experimental method is still required.

3 Cell characterization

It is clear that there is a vast range of topics which have already been, or await to be, examined using isolated ventricular cells. Since it has been emphasized above that it is important that cell characteristics be examined in the general context of the mechanisms under study, careful and critical control experiments are essential for defining cell viability. We have characterized rat myocytes using a variety of criteria, in addition to calcium tolerance and exclusion of stains, one of which has been a careful examination of myocyte morphology and ultrastructure. This has been done with light microscopy (4, 9, 18), scanning electron microscopy (14), thin-section transmission electron microscopy (4, 9, 18, 19) and freeze-fracture techniques (18, 19). Such studies conclude that the population of rod-shaped myocytes are well-preserved ultrastructurally, closely resembling their counterparts in the intact myocardium.

One area of functional studies which has advanced at great speed is that of cardiac electrophysiology, which is considered elsewhere in this volume (13). Work in the London laboratories using suspensions of rat myocytes in high yield have resulted in data describing the efflux of radio-calcium from isolated cells (2), the two-site binding of ouabain to cells in relation to the low-dose inotropic effects of glycosides (1) and the responses of intact cells or cells made hyperpermeable by digitonin to allow assessment of mitochondrial buffering capacity in isolated myocytes (6–8). These results have enabled isolated cells to be characterized using a variety of criteria, and provided sufficient data to indicate that the myocytes have not been functionally compromised by the preparative techniques.

4 Conclusions

It should not be inferred from this brief summary that isolated myocyte research is confined to ventricular cells of the rat or guinea-pig (and only to London). Space limitations preclude discussion of the many reports now appearing in the literature describing the ultrastructure and function of cells isolated from every location of the hearts of a wide range of animals, including the rat. Initial work has also shown that it is certainly possible to obtain cells from human myocardium (15), thus providing a direct link to studies concerned with situations of direct clinical relevance in man. It is to be anticipated that isolated heart cell research will proceed apace over a broad front for many years to come, in the long task of eventually understanding the mechanisms underlying diseased states in the myocardium.

References

1. Adams RJ, Schwartz A, Grupp G, Grupp I, Lee SW, Wallick ET, Powell T, Twist VW, Gathiram P (1982): High-affinity ouabain binding site and low dose positive inotropic effect in rat myocardium. Nature 296: 167–169
2. Baker CJ, Powell T, Twist VW, Williams BC (1981): Calcium efflux from isolated rat ventricular muscle cells. J Physiol Lond 312: 36P
3. Berry MN, Friend DS, Scheuer J (1970): Morphology and metabolism of intact muscle cells isolated from adult rat heart. Circ Res 26: 679–687
4. Dow JW, Harding NGL, Powell T (1981): Isolated cardiac myocytes: I. Preparation of adult myocytes and their homology with the intact tissue. Cardiovasc Res 15: 483–514
5. Dow JW, Harding NGL, Powell T (1981): Isolated cardiac myocytes: II. Functional aspects of mature cells. Cardiovasc Res 15: 549–579
6. Fry CH, Powell T, Twist VW (1981): Net calcium fluxes in isolated rat ventricular muscle cells. J Physiol Lond 315: 17–18P
7. Fry CH, Powell T, Twist VW, Ward JPT (1983): The effects of Na on net calcium ion exchange by digitonin-treated rat cardiac ventricular myocytes. J Physiol Lond 339: 27P
8. Fry CH, Powell T, Twist VW, Ward JPT (1984): Alterations to calcium exchange in digitonin-treated rat ventricular myocytes in response to changes in suspension pH. J Physiol Lond 346: 78P
9. Gould RP, Powell T (1972): Intact isolated muscle cells from the adult rat heart. J Physiol Lond 225: 16–19P
10. Kono T (1969): Roles of collagenases and other proteolytic enzymes in the dispersal of animal tissues. Biochem Biophys Acta 178: 397–400
11. Powell T (1979): Isolation of cells from adult mammalian myocardium. J Mol Cell Cardiol 11: 511–513
12. Powell T (1983): The calcium paradox and isolated myocytes. Eur Heart J 4 (suppl H): 105–111
13. Powell T (1984): Electrophysiology of isolated ventricular myocytes. Basic Res Cardiol (This volume)
14. Powell T, Steen EM, Twist VW, Woolf N (1978): Surface characteristics of cells isolated from adult rat ventricular myocardium. J Mol Cell Cardiol 10: 287–292
15. Powell T, Sturridge MF, Suvarna SK, Terrar DA, Twist VW (1981): Intact individual heart cells isolated from human ventricular tissue. Br Med J 283: 1013–1015

16. Powell T, Twist VW (1976): A rapid technique for the isolation and purification of adult cardiac muscle cells having respiratory control and a tolerance to calcium. Biochem Biophys Res Commun 72: 327–333
17. Powell T, Twist VW (1976): Isoprenaline stimulation of cyclic AMP production by isolated cells from adult rat myocardium. Biochem Biophys Res Commun 72: 1218–1225
18. Severs NJ, Slade AM, Powell T, Twist VW, Warren RL (1982): Correlation of ultrastructure and function in calcium tolerant myocytes isolated from the adult rat heart. J Ultrastruct Res 81: 222–239
19. Slade AM, Severs NJ, Powell T, Twist VW (1983): Isolated calcium-tolerant myocytes and the calcium paradox: an ultrastructural comparison. Eur Heart J 4 (suppl H): 113–122
20. Vahouny GV, Wei R, Starkweather R, Davis C (1970): Preparation of beating heart cells from adult rat. Science 167: 1616–1618

Author's address:

Dr. T. Powell, Department of Medical Physics and Institute of Nuclear Medicine, The Middlesex Hospital Medical School, London, W1P 6 DB, (U.K.)

Ca-tolerant guinea-pig ventricular myocytes as isolated by pronase in the presence of 250 μM free calcium

Z. Bendukidze[1], G. Isenberg[2], and U. Klöckner[2]

[1] A. V. Vishnevsky Surgery Institute, Moscow (USSR)
[2] II. Physiologisches Institut der Universität des Saarlandes, Homburg (F.R.G.)

Summary

A new method to isolate adult cardiocytes with pronase in the presence of 250 μM free calcium was developed. Ultrastructural and electrophysiological properties of these cells were investigated. It is shown that by this method normal calcium-tolerant cells can be obtained.

Key words: isolation procedure, Ca-tolerant myocytes, glycocalyx, action potential

Methodological investigations

We have been able to isolate Ca-tolerant myocytes from the ventricles of adult guinea-pigs by means of the following method: 1. We retrogradely perfuse a "low Ca medium" for 3 min at a rate of 10 ml/min (composition see Table 1). 2. We perfuse 50 ml of an "enzyme medium" containing 100 mg/l pronase (3) (Serva, Heidelberg) and freely ionized calcium (f[Ca]$_o$) at a concentration of 250 μM. We then chop the ventricles and incubate the chunks in fresh enzyme medium for 3 periods of 10 min each (35 °C). 4. We pour the released cells either directly into Tyrode solution containing 3.6 mM CaCl$_2$ or into a storage medium containing 1 mM CaCl$_2$ (Table 1). The resulting released rod-shaped myocytes do not undergo the "Ca paradox", and they are "normal" in their ultrastucture and electrophysiology.

Table 1. Composition of the solutions in mmol/l. f[Ca] from measurements with a Ca-electrode.

Tyrode solution	150 NaCl, 5.4 KCl, 3.6 CaCl$_2$, 1.2 MgCl$_2$, 10 glucose, 10 Hepes, adjusted with NaOH to pH 7.4. f[Ca] 3.6 mM.
"Ca-free" Tyrode	Tyrode solution but nominally Ca-free. f[Ca] $= 3\ \mu$M.
Low Ca medium	100 NaCl, 10 KCl, 1.2 KH$_2$PO$_4$, 5 MgSO$_4$, 20 glucose, 50 taurine, 10 Mops adjusted with KOH to pH 6.9 f[Ca] $= 3\ \mu$M.
Enzyme medium	Low Ca medium plus 100 mg/l pronase E (Serva, Heidelberg) and 1 g/l fatty acid free albumin (Sigma A–6003). f[Ca] adjusted with 120 μM CaCl$_2$ to 250 μM.
Storage medium	Low Ca medium plus 1 mM CaCl$_2$. f[Ca] 0.92 mM.

We compare the above method with the previous procedure used to induce Ca-tolerance by pre-incubating the cells in a "Kraftbrühe" (KB-medium, [9, 10]). Both methods start by exposing the tissue to nominally Ca-free media (Ca contamination in the micromolar range), but they do so for different durations. It is necessary to wash out calcium, since enzymatic dissociation without the preceding exposure to low Ca media delivers only broken cells (own observations, [2]). The enzyme medium used thereafter for tissue perfusion contains CaCl$_2$ as originally proposed by Powell et al. (13), however in that method 10-fold lower concentrations are used. The removal of the [Ca]$_o$ probably destabilizes the cell to cell contacts by uncoupling the cells at the interca-

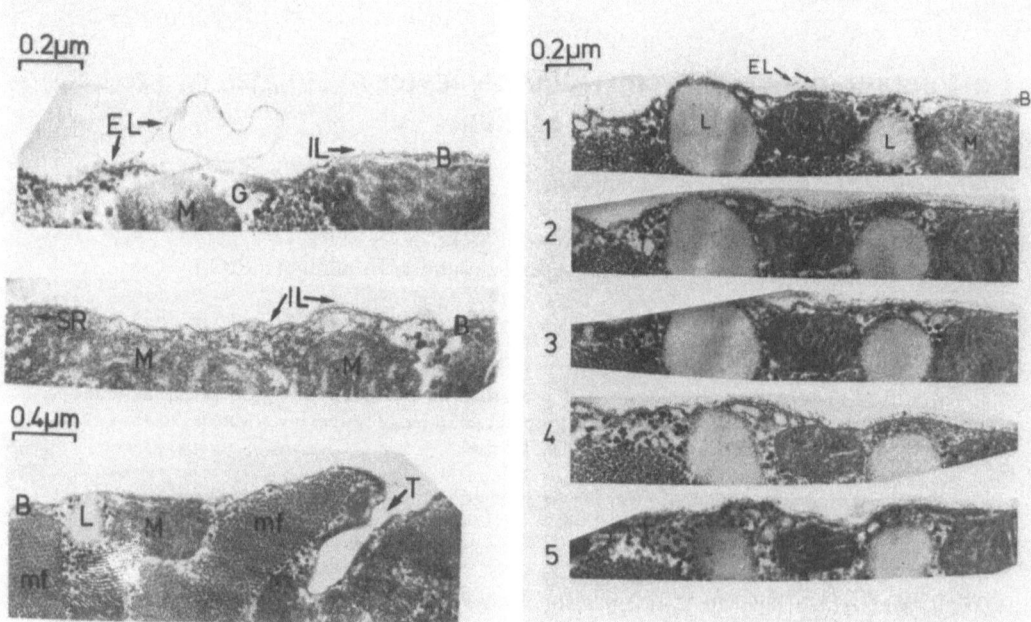

Fig. 1. Low Ca media deteriorate the glycocalix by peeling off the external lamina from the surface coat. The myocytes were chemically fixed by 2% glutaraldehyde plus 1% osmium tetroxide in sodium cacodylate buffer. The glycocalix was stained with 1% ruthenium-red. EL, IL: external lamina and internal lamina (surface coat) of the glycocalix. B: lipid bilayer membrane of the sarcolemma. M: mitochondria. mf: myofilaments. G: glycogen granulae. L: droplets of lipids. Left top: The external lamina forms a "bleb" adjacent to the sarcolemma. Left middle: Absence of the external lamina is suggested by the thin layer of electron-dense material. Left bottom: The external lamina is peeled off the membrane of the transverse tubule. Right: Serial sections (thickness about 80 nm) demonstrate the cleavage of the external lamina from the surface coat.

lated discs (12). But, it also unfortunately induces a leakiness in the surface membrane; myocytes isolated and incubated in low Ca media lose potassium and gain sodium (1, 10), they accumulate calcium in the cytosol (10) and in the mitochondria (4) and they become depleted of ATP (7). Treatment of such cells with "KB-medium" was supposed to alleviate these disturbances (9, 10) whereas with the new method those effects are minimized by shortening the exposure time to the low Ca medium as much as possible. What causes the "hyperpermeability" of the surface membrane? The extensive work of Langer's group has shown that low Ca media deteriorate the glycocalix as the external lamina peels away from the surface coat (cf. ref. [11]). Though we can now extend this observation to the isolated myocytes (Fig. 1, also [9]), we do not agree with the interpretation that the deterioration of the glycocalix is functionally related to the appearance of hyperpermeability. Isolated myocytes taken from Tyrode solution and placed into nominally Ca-free solutions for 20 min have resting potentials of -74 ± 5 mV (mean $\pm$ S. D., n = 10), input resistances 3 times greater than those found in Tyrode solution (100 ± 12 MOhm from voltage clamp measurements between -70 and -80 mV), and they respond to stimuli with 100–200 ms long action potentials (Fig. 2). This result – isolated Ca-tolerant myocytes which do not become

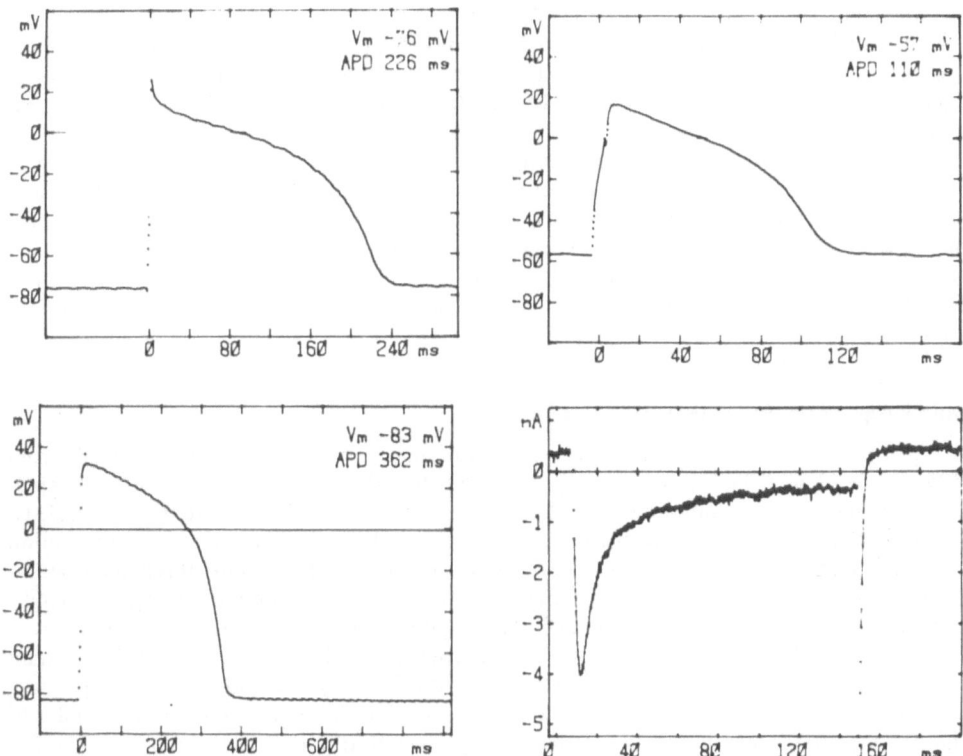

Fig. 2. Action potentials and membrane currents of isolated guinea-pig ventricular myocytes. Top: The cells were superfused with Ca-free Tyrode solution (left) or with the low Ca medium (right) for 20 min. Both solutions were contaminated with 3 μM freely ionized calcium (see Table 1). Below: The cells were superfused with Tyrode solution containing 3.6 mM $CaCl_2$. The membrane currents (right) were recorded during a 160 ms long clamp step from -45 mV to $+5$ mV.

"leaky" in low Ca media – contrasts with our earlier data on myocytes isolated and investigated in media containing less than 10 μM $f[Ca]_o$. Such cells had resting potentials of only -4 mV and input resistances of 1 MOhm, suggesting leakiness (10). This could mean that low calcium media induce "hyperpermeability" only in multicellular tissue, when they split the cell-to-cell contacts. Therefore, we would like to attribute the "hyperpermeability" to membrane lesions (8) which heal over when f [Ca])$_o$ is 250 μM. (In damaged multicellular heart tissue the healing over of lesions crucially depends on f [Ca]$_o$ [5]).

With the notion that $f[Ca]_o$-dependent "healing over" processes stop the leakiness, we need to know for what period of time we must expose the tissue to the low Ca medium. By means of a Ca electrode we have measured $f[Ca]_o$ in the effluent from the Langendorff heart. On perfusing the low Ca medium at a rate of 10 ml/min, the $f[Ca]_o$ fell to 12 μM within the first min, to 6 μM during the second min, and later stabilized at 3 μM (3 μM is the Ca contamination of the nominally Ca-free solutions, see Table 1). A 3 min perfusion by the low Ca medium was sufficient and necessary to condition the heart for the successful enzymatic treatment.

How does the presence of 250 μM $f[Ca]_o$ in the enzyme medium modify the isolation process? Pronase-like collagenase is activated by calcium ions. In order to obtain a similar number of released cells, we need at 9 μM $f[Ca]_o$ about 1.5 g/l protease (Sigma P5380), at 40 μM $f[Ca]_o$

500 mg/l and at 250 μM $f[Ca]_o$ only 100 mg/l. The use of 400 μM $f[Ca]_o$ and 100 mg/l protease does not further increase the cell release, and in 1 mM $f[Ca]_o$ the tissue dissociation is inhibited. (Some samples of the crude pronase are strongly contaminated by calcium; 100 mg pronase E from Serva, Heidelberg, contain 120 μM $f[Ca]_o$, 100 mg protease Sigma P6141 contain 33 μM, and 100 mg of protease Sigma P5380 contain 6 μM $f[Ca]_o$).

Why use pronase instead of collagenase? Firstly, the dissociation in 250 μM $f[Ca]_o$ works just as well when collagenase is used instead of pronase. For comparison, we repeated the method given above but added 100 mg/l collagenase (Sigma C-2139) and 20 mg/l trypsin instead of pronase to the "enzyme medium" (250 μM $f[Ca]_o$ other constituents identical). This procedure delivered a similar number of Ca-tolerant cells whose electophysiological properties did not differ from those obtained with pronase (3 hearts). The main reason for preferring pronase is that it is about 10 times cheaper.

Morphology of the myocytes in Tyrode solution

When we superfused the cells with Tyrode solution at 35 °C the increase of $f[Ca]_o$ from 0.25 mM to 3.6 mM did not reduce the yield of rod-shaped cells (between 50–70%). The myocytes were ungranulated and showed a clear striation pattern. The electromicrographs revealed a well preserved intracellular ultrastructure. However, the glycocalix showed some deterioration because its external lamina was peeled off from the surface coat. To quantify this damage we investigated the surface membrane with serial sections over a total length of 2000 μm (magnification 120000; 5 cells): in 54% of the surface lining, the external lamina was peeled off to form fragments or blebs, in 23% the lipid bilayer was covered by electron-dense material less than 25 nm thick, and only in the remaining 23% was a layer thicker than 25 nm apparent, suggesting that the external lamina was still adherent. The lining of the T-tubules shows ratios similar to those of the surface membrane. In summary, the perfusion with low Ca medium for 3 min followed by enzymatic treatment caused a deterioration of the glycocalix similar to that found in earlier work (8, 9).

Electrophysiology

By means of either conventional microelectrodes or patch electrodes, we measured resting potentials of -82 ± 3 mV (mean $\pm$ S. D., n = 25). The action potentials had an amplitude of 115 ± 15 mV. When we clamped the cells to a holding potential of -45 mV, an outward current of 0.35 nA was measured (see Fig. 2). A step to $+5$ mV evoked a negative deflection due to the calcium inward current (i_{Ca}) which peaked to -3.6 nA (negative current surge minus current at the end of the 160 ms long pulse). On repolarization back to -85 mV, a short lasting inwardly directed tail current appeared. These rough qualitative data suggest that the myocytes isolated with the pronase medium have "normal" electrophysiological properties.

References

1. Altschuld R, Gibb L, Ansel A, Hohl C, Kruger FA, Brierly CP (1980): Calcium tolerance of isolated rat heart cells. J Mol Cell Cardiol 12: 1383–1395
2. Berry MN, Friend DS, Scheuer J (1980): Morphology and metabolism of intact muscle cells isolated from adult rat heart. Circ Res 26: 679–687
3. Bustamante JQ, Watanabe T, McDonald TF (1982): Non-specific proteases: a new approach to the isolation of adult cardiocytes. Can J Physiol Pharmacol 59: 906–910
4. Chiesi M, Ho MM, Inesi G, Somlyo AV, Somlyo AP (1981): Primary role of sarcoplasmic reticulum in phasic contractile activation of cardiac myocytes with shunted myolemma. J Cell Biol 91: 728–742

5. Deleze J (1970): The recovery of resting potential and input resistance in sheep heart injured by knife or laser. J Physiol 208: 547–562
6. Dow JW, Harding NGL, Powell T (1981): Isolated cardiac myocytes. I. Preparation of adult myocytes and their homology with the intact tissue. Cardiovasc Res 15: 483–514
7. Farmer BB, Harris RA, Jolly WW, Hathaway DR, Katzberg , Watanabe AM, Whitlow AL, Besch HR (1977): Isolation and characterization of adult rat heart cells. Arch Biochem Biophys 179: 449–458
8. Fry DM, Scales D, Inesi G (1979): The ultrastructure of membrane alterations of enzymatically dissociated cardiac myocytes. J Mol Cell Cardiol 11: 1151–1163
9. Isenberg G, Klöckner U (1980): Glycocalix is not required for slow inward calcium current in isolated rat heart myocytes. Nature 284: 358–360
10. Isenberg G, Klöckner U (1982): Calcium tolerant ventricular myocytes prepared by preincubation in a "KB-medium". Pflügers Arch 395: 6–18
11. Langer GA, Frank JS, Philipson KD (1982): Ultrastructure and calcium exchange of the sarcolemma, sarcoplasmic reticulum and mitochondria of the myocardium. Pharmac Ther 16: 331–376
12. Muir A (1966): The effects of divalent cations on the ultrastructure of the perfused rat heart. J Anat 101: 239–261
13. Powell T, Terrar DA, Twist VW (1980): Electrical properties of individual cells isolated from adult rat ventricular myocardium. J Physiol 302: 131–153

Author's address:

Dr. G. Isenberg, II. Physiologisches Institut der Universität des Saarlandes, 6650 Homburg/Saar (F.R.G.)

Guinea-pig atrial cardioballs*)

M. Bechem, F. Pieper and L. Pott

Institut für Zellphysiologie der Ruhr-Universität Bochum, Bochum (F.R.G.)

Summary

Myocytes from atria of adult guinea-pigs were isolated by means of a previously described enzyme perfusion with some modifications.

The problem of Ca-intolerance of the dispersed cells was circumvented by (i) avoiding cooling of the cells below 25 °C and (ii) increasing the Ca concentration slowly already during the enzyme perfusion.

Isolated atrial myocytes were taken in long-term cell culture. Under this condition they become spherical within about 2 days. The rounded stage (cardioballs), which is found in the cultures for a period of ca. 10 days is highly suited for electrophysiological studies using the different recording configurations of the patch clamp technique, including 'tight-seal whole-cell recording' with simultaneous cell dialysis.

Key words: single heart cell, cardioball, cell culture, whole-cell recording

Introduction

Enzymatic dispersion of Ca-tolerant ventricular myocytes for electrophysiological and metabolic investigations has been becoming routine in many laboratories during the past few years (13, 14, 8, 11: for further literature see 6, 16). Only a few reports so far are concerned with mammalian atrial myocytes (2, 3, 5). In this report we describe the technique of isolation and culturing of atrial myocytes from hearts of adult guinea-pigs. The major advantage of this experimental approach is that in culture these cells pass through a developmental stage which is highly suited for electrophysiological studies using recently developed techniques (7, 10).

Methods

The method for dispersion of atrial cells is similar to that used by other authors for ventricular tissue (6, 16), and has been described in detail recently (5). The major modifications with regard to most of the previous reports are: (i) a mixture of collagenase and elastase is used for the enzymatic dispersion of the tissue (collagenase: Sigma 2139, 1 mg/ml; elastase: Serva 20391, 10 µl suspension/ml); (ii) during the enzyme perfusion, which lasts for 30–60 min, the Ca concentration in the recirculating solution (30 ml) is increased stepwise by adding 0.3 ml of culture medium once every 10 minutes. This means, that at the end of the enzyme perfusion the Ca concentration is in the order of magnitude of 10^{-4}M. The myocytes obtained by this method are Ca-tolerant without further precautions, provided the temperature is kept above 25 °C throughout (cf. ref. 1).

Results and discussion

A yield of up to 80% of intact, i. e. spindle shaped, cross-striated quiescent myocytes (Fig. 1a) can be obtained by the method described. When taken in culture, the morphology of the cells

---

*) This work was supported by a grant from the Ministerium für Wissenschaft und Forschung des Landes Nordrhein-Westfalen

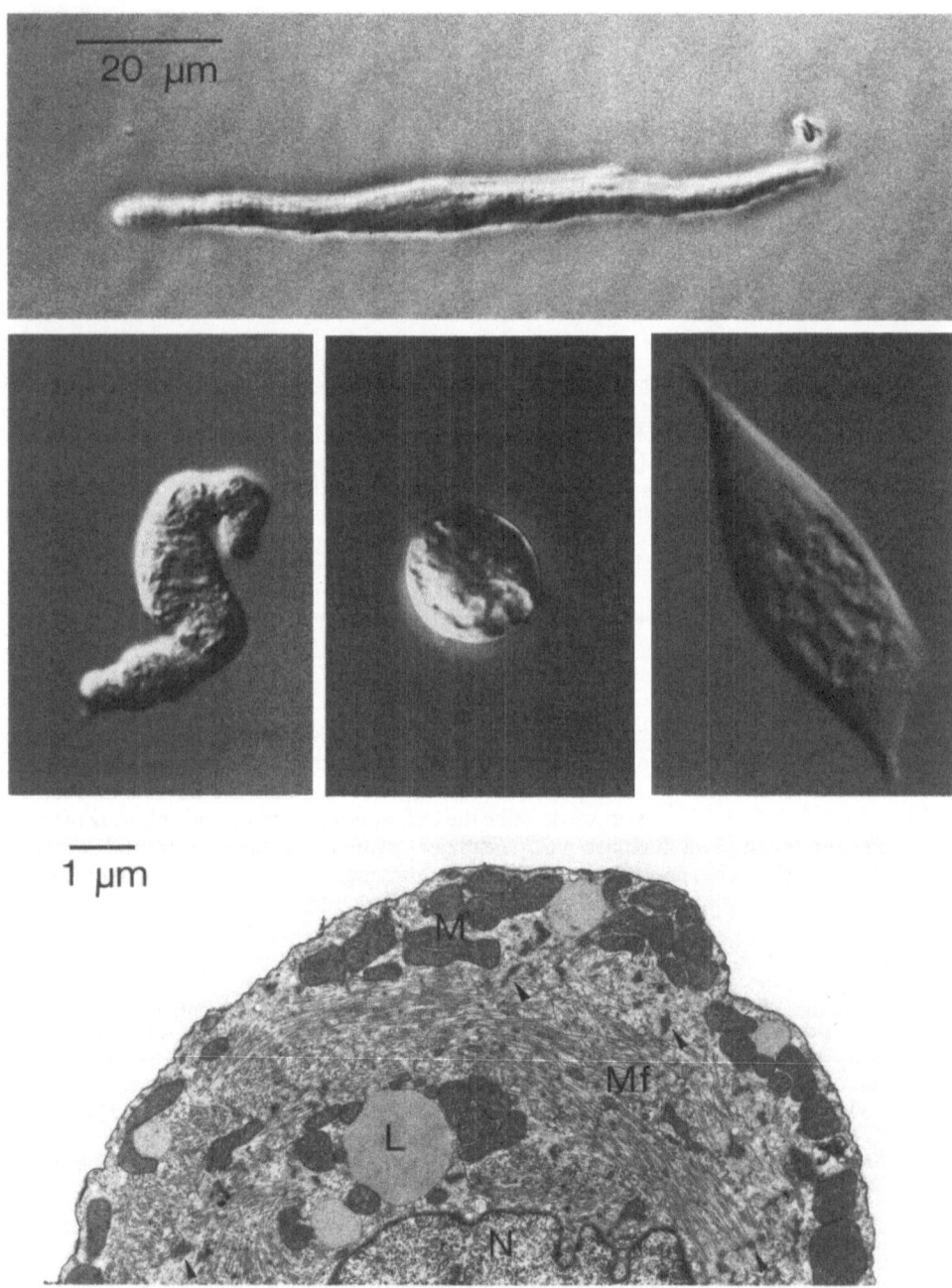

Fig. 1. Developmental stages of atrial myocytes in culture.
Top to bottom, left to right: a. Atrial myocyte immediately after enzymatic isolation. b. Shortened myocyte after 22 h in culture. c. Atrial cardioball after 3 days in culture. d. Flattened myocyte photographed after 11 days. Cells were grown on glass coverslips in order to render possible the use of interference phase contrast optics. e. Low magnification electron micrograph of a myocyte in the cardioball stage. Note that myofilaments (Mf) appear in cross- and longitudinal section. L: Lipid droplet, M: Mitochondria, N: Nucleus. The arrowheads denote Z-disc material.

changes following a characteristic pattern. The first visible structural alteration is the disappearance of cross striation, which occurs after a few hours. This is followed by a progressive shortening (Fig. 1b). After about 2 days most of the cells in a culture dish are completely rounded and attach to the substratum (Fig. 1c). This 'cardioball-stage' is maintained for several days and is followed by a progressive flattening. After a period of 10 days in culture a fraction of ca. 20% of the myocytes is still spherical, whereas the majority is flattened (Fig. 1d).

The morphological changes visible on the light microscopic level are accompanied by substantial alterations of the internal ultrastucture. As shown on the electron-micrograph in Fig. 1e the organization of the contractile proteins into sarcomere units is completely lost. Actin and myosin filaments, which can still be identified, traverse the cell without any detectable order, and Z-disc material can be found distributed irregularly in the cytoplasm. The membrane of the cardioballs appears smooth, and invaginations were never detected.

As compared to freshly isolated myocytes the cardioballs provide several advantages, at least with regard to electrophysiological studies. One practical advantage is, that once isolated, cells can simply be taken 'off the shelf' for more than a week.

The second advantage results from the fact that the cells adhere firmly to the bottom of the culture dish. This facilitates the isolation of cell-free membrane patches for measurements of single ionic channel current events (7, 3, 5). The principal advantage, however, is the size and the geometry of atrial cardioballs. The smallest cells are about 15 μm in diameter, which by calculation yields a membrane surface of about 700 μm^2 (the 'capacitive surface' is about 20% higher). As compared to freshly isolated atrial myocytes the surface is reduced by a factor of at least 3. Ventricular myocytes, which are being used for investigations of membrane currents (e. g. 9, 15) possess a surface area which is larger by more than one order of magnitude.

Cells of the size described here are highly suited for measurements using the whole cell recording configuration of the patch clamp technique (10), which allows measurements of membrane currents under control of both the extracellular and the intracellular environment. By means of this technique we have so far identified the rapid Na^+ inward current, two types of K^+ currents, one of which is modulated by acetylcholine via muscarinic receptors (12) whereas the second type seems to be modulated by the intracellular ATP concentration (4). A study of I_{Ca} will be presented in an accompanying paper (Bechem & Pott, this volume).

References

1. Altschuld R, Gibb L, Ansel A, Kruger FA, Brierley GP (1980): Calcium tolerance of isolated rat heart cells. J Mol Cell Cardiol 12: 1383–1395
2. Belardinelli L, Isenberg G (1983): Isolated atrial myocytes: adenosine and acetylcholine increase potassium conductance. Am J Physiol 244: H734–737
3. Bechem M, Glitsch HG, Pott L (1983a): Properties of an inward rectifying K channel in the membrane of guinea-pig atrial cardioballs. Pflügers Arch 399: 186–193
4. Bechem M, Pott L (1984): K-channels activated by loss of intracellular ATP in guinea-pig atrial cardioballs. J Physiol (London) 348: 50P
5. Bechem M, Pott L, Rennebaum H (1983b): Atrial muscle cells from hearts of adult guinea-pigs in culture: a new preparation for cardiac cellular electrophysiology. Europ J Cell Biol 31: 366–369
6. Dow JW, Harding NGL, Powell T (1981): Isolated cardiac myocytes. I. Preparation of adult myocytes and their homology with the intact tissue. Cardiovasc Res 15: 483–514
7. Hamill OP, Marty A, Neher E, Sakmann B, Sigworth F (1981): Improved patch-clamp techniques for high-resolution current recording from cells and cell-free membrane patches. Pflügers Arch 391: 85–100
8. Isenberg G, Klöckner U (1982a): Calcium tolerant ventricular myocytes delivered by pre-incubation in a "KB-medium". Pflügers Arch 395: 6–18
9. Isenberg G, Klöckner U (1982b): Calcium currents of isolated bovine ventricular myocytes are fast and of large amplitude. Pflügers Arch 395: 30–41

10. Marty A, Neher E (1983): Tight-seal whole-cell recording. In: Sakmann B, Neher E (eds). Single channel recording. Chapter 7, Plenum, New York, 107–122
11. Piper HM, Probst J, Schwartz P, Hütter FJ, Spieckermann PG: Culturing of calcium stable adult cardiac myocytes. J Moll Cell Cardiol 14: 397–412
12. Pott L, Bechem M (1984): Properties of muscarinic K channels in guinea-pig atrial cardioballs as studied by ACh injection into patch clamp pipettes. Pflügers Arch (abstr) in press
13. Powell T, Twist VW (1976): A rapid technique for the isolation and purification of adult cardiac muscle cells having respiratory control and a tolerance to calcium. Biochem Biophys Res Commun 72: 327–333
14. Powell T, Terrar DA, Twist VW (1980): Electrical properties of individual cells isolated from adult rat ventricular myocardium. J Physiol (London) 302: 131–153
15. Sakmann B, Trube G (1984): Conductance properties of single inwardly rectifying potassium channels in ventricular cells from guinea-pig heart. J Physiol (London) 347: 641–657
16. Trube G (1983): Enzymatic dispersion of heart and other tissues. In: Sakmann B, Neher E (eds). Single channel recording, Chapter 4, Plenum, New York, 69–76

Author's address:

Dr. Martin Bechem, Bayer Forschungszentrum, Institut für Pharmakologie, 5600 Wuppertal (F.R.G.)

Determination of isolated myocyte viability: Staining methods and functional criteria

J. Y. Cheung, A. Leaf and J. V. Bonventre

Department of Medicine, Massachusetts General Hospital and Harvard Medical School, Boston, Massachusetts (U.S.A.)

Summary

Single ventricular cells tolerant to physiological levels of calcium have been successfully isolated from adult rat heart. These cells exhibit morphological as well as functional characteristics of the intact myocardium. Under anaerobic incubations either in the absence of extracellular calcium or in the presence of 6% PEG, myocytes excluded Trypan blue despite severe derangements in biochemical functions. We suggest that Trypan blue entry may be a rather insensitive criterion of cell injury. We recommend the combination of rod-shape morphology, ATP levels and the ability to contract under external pacing as alternative simple tests of cell viability.

Key words: viability criteria, Trypan blue, adenosine triphosphate, anoxia, calcium

Introduction

Cardiac myocytes isolated from adult rat ventricles have proven to be a useful tool in the delineation of normal cardiac structure and function at the molecular level. However, one of the major problems in the study of single cells is the definition of a "viable" cell, or conversely, cell death. The present paper addresses staining and functional criteria of viability as they relate to the isolated cardiac myocyte.

Methods

Ca^{++}-tolerant cardiac myocytes were isolated from adult rat ventricles as described in detail elsewhere (5). Cells were incubated in Krebs-Henseleit bicarbonate buffer containing 2% bovine serum albumin and 5 mM pyruvate. When indicated, either calcium (normally 1.25 mM) was omitted from the medium ($Ca^{++} < 10$ μM) or 6% polyethylene glycol (PEG MW 7000–8000) was added. Cell suspensions were equilibrated with humidified 95% O_2-5% (O_2 cells) or 95% N_2-5% CO_2 (N_2 cells) and incubated in 25 ml siliconized Erlenmeyer flasks in a shaking water bath maintained at 37 °C and 100 cycles/min. Immediately prior to the start and after 45 min of incubation, aliquots of cell suspension were counted on a bright-line hemacytometer for rod-shaped, round and Trypan blue negative cells. Cellular ATP levels and ionic contents were determined at 45 min. Rates of ^{14}C-phenylalanine incorporation into protein were measured during 2 hours of recovery from the initial anoxic incubation (5). Creatine phosphokinase (CPK) released into the medium was measured using the SIGMA UV-46 assay system. Aliquots of cells were also prepared for examination by electron microscopy (5).

Results

Structural and functional characteristics of isolated cardiac myocytes

Myocytes isolated from adult rat ventricles existed as individual cells, greater than 80 to 85% of which excluded Trypan blue stain (Fig. 1A). Approximately 85 to 90% of Trypan blue negative cells maintained rod-shaped morphology (Fig. 1B) as well as cross striations. The ultrastruc-

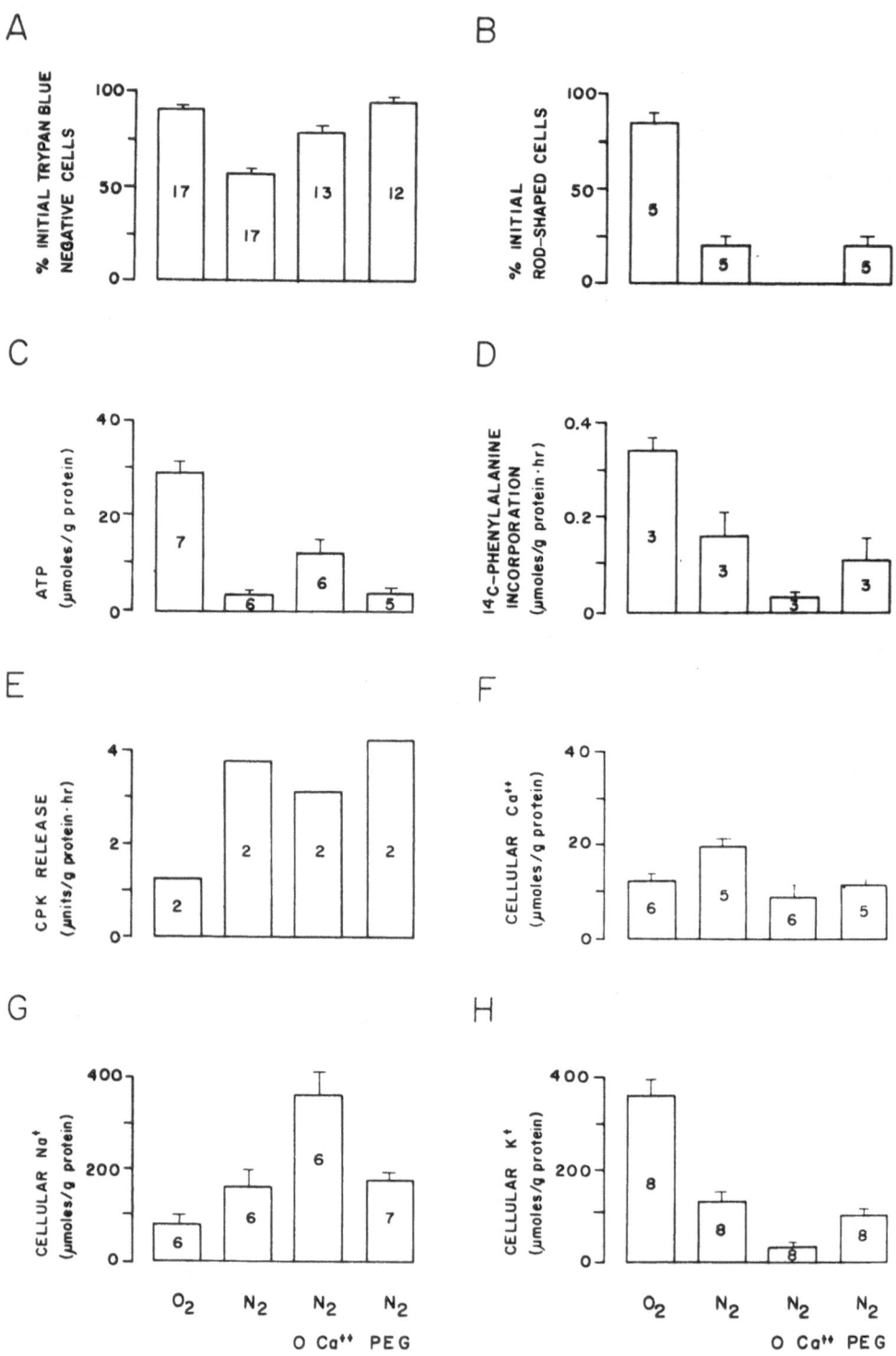

Fig. 1. Correlation among Trypan blue exclusion test, cell morphology and functional criteria of cell viability. See text for details. In Fig. 1B, cells in anoxic, Ca^{++}-free medium assumed a rectangular structure and could not be classified as rod-shaped or round cells. In Fig. 1D, rates of ^{14}C-phenylalanine incorporation into protein were measured during 2 hours of aerobic incubation following the initial 45 min incubation period.

tural appearance of rod-shaped myocytes was indistinguishable from that of cells in intact cardiac muscle (Fig. 2). Functionally, aerobic myocytes maintained normal ATP levels (Fig. 1C) when compared to the isolated perfused rat heart (13) although the rate of protein synthesis (Fig. 1D) was approximately $1/2$ that of the intact organ (12). The latter difference may be due to the absence of cardiac work in isolated myocytes (8). Intracellular ionic determinations showed that normal Na^+ and K^+ gradients were maintained (Figs. 1G and 1H). When placed in a Northtrup-Kunitz microcataphoresis glass chamber and periodically stimulated with a pulse generator, rod-shaped myocytes were observed to contract synchronously and repeatedly. In addition, contraction could be blocked completely with either 1 µM of verapamil or nifedipine suggesting that slow calcium channels were intact (4). Thus both structural and functional considerations suggested that a vast majority of myocytes were viable after the isolation procedure.

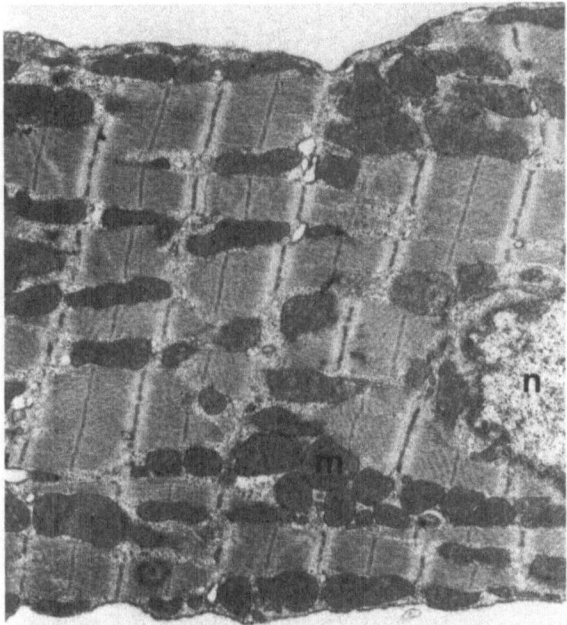

Fig. 2. Transmission electron micrograph of an isolated myocyte incubated in oxygenated Krebs-Henseleit bicarbonate buffer. The banding pattern and the ultrastructure of the mitochondria (M) and nucleus (N) appear normal. 9300X.

Effect of anoxia on structure and function of isolated cardiac myocytes

After 45 min of anaerobic incubation, approximately half of the myocytes permitted Trypan blue entry, indicating damage to the plasma membrane (Fig. 1A). Most of the cells lost their rod-shaped configuration (Fig. 1B). Most of the cells lost their ability to contract to external electrical stimuli.

Ultrastructurally contraction bands were evident, mitochondria were swollen with broken cristae and their matrices appeared washed out (Fig. 3). Functional criteria of cell viability also reflected severe cell injury: depressed ATP levels and rates of protein synthesis, enhanced CPK leakage, and altered cellular ionic composition (Fig. 1C to 1H). Under these conditions, there was a close correlation among cell morphology, biochemical function and vital dye exclusion.

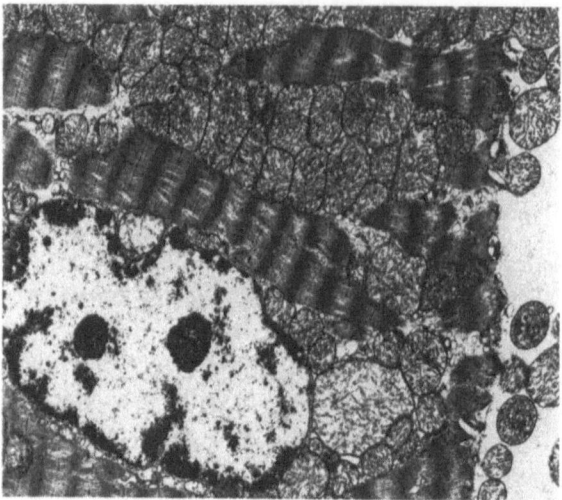

Fig. 3. Transmission electron micrograph of an isolated myocyte incubated in anaerobic medium containing 1.25 mM calcium. The cell is contracted as seen by the considerable overlap of the actin and myosin filaments. Mitochondria are swollen with broken cristae and the inner matrix appears washed out. 6000X.

The observation that approximately half of the rounded cells excluded Trypan blue despite loss of synthetic and contractile functions suggests that uptake of vital dyes may be a late manifestation of cell injury.

Effect of extracellular calcium removal on anaerobic myocyte survival

Accumulation of calcium in the cell has been observed in hearts irreversibly injured by ischemia (17). Recently, it has been proposed that influx of extracellular calcium is the final common pathway that caused cell death from various toxic agents (16).

Thus it is of interest to evaluate the role of extracellular calcium on anoxic cell injury. After 45 min of anaerobic incubation in calcium-free buffer, myocytes were able to exclude Trypan blue (Fig. 1A) and became more rectangular in shape (decrease in length-to-width ratio). Ultrastructurally, these myocytes appeared intact as evidenced by relaxed myofibrils and normal mitochondrial morphology (Fig. 4). Thus by staining technique as well as ultrastructural examination, these cells appeared "viable". By contrast, metabolic activities were grossly deranged as indicated by low ATP levels (Fig. 1C), rates of protein synthesis (Fig. 1D) and severely altered cellular ionic compositions (Fig. 1G and 1H). It is interesting that release of CPK was similar in anaerobic cells whether they were incubated in 1.25 mM or 0 mM Ca^{++} buffer (Fig. 1E) despite "intact" plasma membrane by Trypan blue exclusion criterion. Subsequent introduction of substrates, O_2 and calcium in attempts to correct the biochemical abnormalities was precluded by the fact that cardiac cells previously exposed to calcium-free environment are damaged by massive calcium influx on reintroduction to calcium-containing medium, the well known "calcium paradox" (19).

Effect of polyethylene glycol on anaerobic myocyte survival

Non-permanent solutes such as mannitol and polyethylene glycol have been found to protect the heart (18) and kidney (7) from acute ischemic injury. However, it has been difficult to establish a direct cellular protective effect from *in vivo* studies because of potential circulatory or other

Fig. 4. Transmission electron micrograph of an isolated myocyte incubated in anaerobic calcium-free medium. The cells retain their ultrastructural integrity and are comparable in appearance to those cells incubated in normal oxygenated medium. 12900X.

effects of these compounds. Recently it has been reported that polyethylene glycol prevented anoxic cultured renal epithelial cells from taking up Trypan blue, suggesting that protection was afforded at the cellular level (11). In isolated myocytes, polyethylene glycol also prevented Trypan blue uptake after 45 minutes of anoxia (Fig. 1A). However, a more detailed examination showed that PEG-treated myocytes incubated anaerobically lost their rod-shaped morphology (Fig. 1B) and had biochemical parameters indistinguishable from anoxic cells not treated with PEG. (Fig. 1C to 1H). Thus, ATP levels fell, CPK release was enhanced, cell K^+ decreased, cell Na^+ increased during anoxia, and amino acid incorporation into protein was impaired during the postanoxic "recovery period". In addition, recovery of ATP levels after the anoxic insult was not enhanced in PEG-treated cells (14.0 ± 4.2 and 12.2 ± 0.9 μ moles/g protein in N_2 and N_2-PEG cells, respectively compared to 48.5 ± 15.5 μ moles/g protein in O_2 cells; measured 2 hours after return of O_2 and substrates including 100 μM of adenosine). Moreover, on removal of PEG, anaerobic myocytes immediately permitted Trypan blue entry. It should be noted that myocytes, already damaged by the isolation procedure or by heat permitted entry of Trypan blue even in the presence of PEG. Despite the discrepancy between the staining and functional criteria of cell viability, it is reasonable to conclude that PEG exerts no beneficial effect on anoxic myocytes.

Discussion

Many techniques have been employed to determine cell viability. As is often the case the many techniques belie the fact that none is completely acceptable. Since mature cardiac myocytes do not divide, reproductive capability – the standard criterion of cell viability – cannot be used as a test of viability. The single most frequently used test of viability is vital dye (e.g. Trypan blue) exclusion. Cells that are injured and thus have leaky plasma membranes readily take up the dye and are easily distinguished from normal cells. Other viability tests which also focus on plasma membrane integrity include fluorescein (15), and cellular enzyme release into the medium (1,2).

Functional criteria of cell viability include membrane potential (2), cellular Na^+ and K^+ concentrations (2), cellular surface charge density (14), amino acid uptake (6), DNA and protein synthesis (2,3) ATP levels (9), and cellular calcium contents (10). The difficulty with these techniques is that each measures only one aspect of the many functions of a living cell. In addition, some of these vital cellular functions can be simulated in the test-tube with "lifeless" systems, e.g. amino acid uptake in membrane vesicles, protein synthesis in polyribosomal systems. In general, vital dye staining correlated well with other viability criteria. In isolated myocytes, however, we have clearly demonstrated two conditions in which there is disagreement between Trypan blue exclusion and functional characteristics, i.e., anaerobic incubation in Ca^{2+}-free medium or in the presence of a poorly penetrating solute, PEG. In this light, it is interesting to note that hypertonically-treated baby hamster kidney cells, although permeable to Trypan blue stain, synthesized DNA, RNA and protein rapidly when supplied with appropriate substrates and cofactors. In addition, cells previously exposed to hypertonic medium, when returned to isotonic medium, were able to "reseal", exclude Trypan blue, and resumed the same growth rate as untreated cells (3). Baur et al. (2) also found that isolated liver cells can be reversibly stained with Trypan blue by decreasing the pH of the cellular bathing medium. These studies on kidney and liver cells as well as ours on isolated myocytes clearly stress the need to correlate Trypan blue exclusion with functional criteria of viability. Our experience with isolated myocytes suggests that rod-shaped morphology, ATP levels and ability to contract in response to electrical stimuli are more sensitive indicators of cellular well-being than Trypan blue exclusion.

Acknowledgement

This work was supported by National Heart, Lung and Blood Institute Program Project Grant HL-06664 and HL-31513. J. Y. Cheung was supported by National Heart, Lung and Blood Institute National Research Service Award HL-06378. J. V. Bonventre was supported in part by National Institute of Arthritis, Diabetes and Digestive and Kidney Diseases New Investigator Research Award AM-27957.

References

1. Acosta D, Pucketts M, McMillin R (1978): Ischemic myocardial injury in cultured heart cells: leakage of cytoplasmic enzymes from injured cell. In Vitro 14: 728–732
2. Baur H, Kasperek S, Pfaff E (1975): Criteria of viability of isolated liver cells. Hoppe-Seyler Z Physiol Chem 356: 827–838
3. Castellot Jr, JJ, Miller MR, Pardee AB (1979): Animal cell reversibly permeable to small molecules. Proc Natl. Acad Sci USA 75: 351–355
4. Cheung JY, Leaf A, Bonventre JV (1984): Mechanism of protection by verapamil and nifedipine from anoxic injury in isolated cardiac myocytes. Am J Physiol 246: C323–C329
5. Cheung JY, Thompson IG, Bonventre JV (1982): Effects of extracellular calcium removal and anoxia on isolated rat myocytes. Am J Physiol 243: C184–C190
6. Edmondson JW, Bang NU (1981): Deleterious effects of calcium deprivation on freshly isolated hepatocytes. Am J Physiol 241: C3–C8
7. Frega NS, DiBona DR, Leaf A (1979): The protection of renal function from ischemic injury in the rat. Pflügers Arch 481: 159–164
8. Hjalmarson A, Isaksson O (1972): In vitro work load and rat heart metabolism. I. Effect on protein synthesis. Acta Physiol Scand 86: 342–352
9. Jennings RB, Hawkins HK, Lowe JE, Hill ML, Klotman S, Reimer KA (1978): Relation between high energy phosphate and lethal injury in myocardial ischemia in the dog. Am J Pathol 92: 187–214
10. Katz AM, Reuter H (1979): Cellular calcium and cardiac cell death. Am J Cardiol 44: 188–190
11. Kreisberg JI, Mills JW, Jarrell JA, Rabito CA, Leaf A (1980): Protection of cultured renal tubular epithelial cells from anoxic cell swelling and cell death. Proc Natl. Acad Sci USA 77: 5445–5447

12. McKee EE, Cheung JY, Rannels DE, Morgan HE (1978): Measurement of the rate of protein synthesis and compartmentation of heart phenylalanine. J Biol Chem 253: 1030–1040
13. Neely JR, Rovetto MJ, Whitmer JT, Morgan HE (1973): Effects of ischemia on function and metabolism of the isolated working rat heart. Am J Physiol 225: 651–658
14. Pfaff E, Schuler B, Krell H, Hoke H (1980): Viability control and special properties of isolated rat hepatocytes. Arch Toxicol 44: 3–21
15. Rotman B, Papermaster BW (1966): Membrane properties of living mammalian cells as studied by enzymatic hydrolysis of fluorogenic ester. Proc Natl Acad Sci USA 55: 134–141
16. Schanne FAX, Kane AB, Young EE, Farber JL (1979): Calcium dependence of toxic cell death: a final common pathway. Science 206: 700–702
17. Shen AC, Jennings RB (1972): Myocardial calcium and magnesium in acute ischemic injury. Am J Pathol 67: 417–440
18. Willerson JT, Powell Jr JW, Gujiney TE, Stark JJ, Sanders CA, Leaf A (1972): Improvement in myocardial function and coronary blood flow in ischemic myocardium after mannitol. J Clin Invest 51: 2989–2998
19. Zimmerman ANE, Daems W, Hulsmann WC, Snijder J, Weiss E, Durrer D (1967): Morphological changes of heart muscle caused by successive perfusion with calcium-free and calcium-containing solutions. (Calcium paradox). Cardivasc Res 1: 201–209

Author's address:

Joseph Y. Cheung, M.D., Ph.D., Renal Unit, Massachusetts General Hospital, Boston, MA 02114 (U.S.A.)

Sarcolemma-bound calcium. Its importance for cell viability

M. Borgers, F. Thoné, and L. Ver Donck

Laboratory of Cell Biology, Department of Life Sciences, Janssen Pharmaceutica, Beerse (Belgium)

Summary

We report the presence of a sarcolemma-associated Ca^{2+}-pool in the intact myocardium and isolated ventricular myocytes of the rat. Ca^{2+}-deposits, which are visualized in the electron microscope as 20 nm thick particles, could be demonstrated when fixation of the tissue was performed in the presence of high inorganic phosphate and most probably represent stable Ca^{2+}-acidic phospholipid-phosphate complexes. Various pathophysiological conditions all leading to intracellular Ca^{2+}-overload and compromising myocardial cell viability were imposed and the distribution of Ca^{2+} was assessed. It was found that in all these conditions cellular Ca^{2+}-overload was preceded or at least accompanied by a loss of sarcolemma-associated Ca^{2+}-deposits. It is concluded that this Ca^{2+}, possibly bound to acidic phospholipids of the sarcolemmal bilayer, plays a role in the maintenance of sarcolemmal integrity. These cytological observations support a concept previously proposed by Langer and his group on the controlling role of membrane Ca^{2+} in the overall cellular Ca^{2+}-homeostasis.

Key words: sarcolemma, calcium, glycocalyx, Ca paradox, ischemia

Introduction

Apart from its possible participation in one of the events of excitation-contraction coupling (1, 2), sarcolemma-bound Ca^{2+} might play an important role in the maintenance of homeostasis for Ca^{2+} and possibly other cations (3). Although stabilizer Ca^{2+}, associated with negatively charged glycocalyx components, might be involved, the bulk of peripherally-bound Ca^{2+} is within the lipid bilayer itself. Recent work showed the preferential association of sarcolemmal Ca^{2+} with acidic phospholipids (4). According to some authors such Ca^{2+} is most probably located at the cytoplasmic face of the membrane (1, 5). In recent studies (6, 7) the association of Ca^{2+} with the sarcolemma was confirmed ultrastructurally (7). In order to characterize this sarcolemma-bound Ca^{2+}, we assayed its ultrastructural localization after various pathophysiological conditions imposed on either intact cardiac muscle or isolated myocytes, thereby predisposing to intracellular Ca^{2+}-overload and compromising cell viability.

Methods used to localize Ca^{2+}

The method used to localize sarcolemma-bound Ca^{2+}, which has been detailed elsewhere (7, 8), is based on the stabilization of membrane Ca^{2+} by a high concentration of inorganic phosphate (90 mM KH_2PO_4) during the initial fixation of the tissue with 3% glutaraldehyde at room temperature. Acidic phospholipids are likely candidate binding sites for membrane-associated Ca^{2+} since it has been shown that under *in vitro* circumstances a stable complex is formed when Ca^{2+}, acidic phospholipids (not neutral ones) and inorganic phosphate are mixed (9). A similar association of Ca^{2+} with inner membrane components has been recently demonstrated in human red blood cells (8). After glutaraldehyde-phosphate fixation for 2–24 h, 100 μm thick vibratome sections were cut and dropped into a solution of 1% OsO_4 and 2% K-pyroantimonate. The origi-

nal aim of this treatment was to convert the Ca^{2+} complex into an insoluble antimonate salt, however, recent X-ray microanalysis of the sarcolemma deposits revealed no antimonate (De Bruyn, personal communication). Nevertheless, the presence of antimonate during the OSO_4 postfixation procedure favorably influenced the stability of the complex, preventing its partial dissolution during the subsequent preparatory procedure for electron microscopy. Indeed, only a small amount of membrane Ca^{2+} deposits is recovered when antimonate is omitted. However, the deposits are morphologically and microanalytically indistinguishable from those in antimonate-treated preparations.

Results and discussion

Visualization of Ca^{2+} in normal cardiac muscle

In perfusion fixed isolated rat hearts, the Ca^{2+}-deposits were almost exclusively confined to the sarcolemma, transverse tubules and the intercalated disks. At the nexus part of the disk the precipitate was deposited in a typically paired position. Mitochondria were not or only weakly positive. Sarcoplasmic reticulum did not reveal any deposits with this method. This is not surprising since it is very unlikely that the Ca^{2+} residing in this organelle would be membrane (phospholipid) bound. Although difficult to delineate with great precision to which side of the membrane the deposits settle because of their 20 nm size, the majority of deposits bulged towards the cytoplasmic face of the sarcolemma. Therefore, we consider the Ca^{2+} deposits associated with bilayer rather than glycocalyx components. The localization of Ca^{2+} in single cardiac myocytes (Fig. 1), isolated according to Piper et al. (10), closely resembled that of the intact heart. The observation

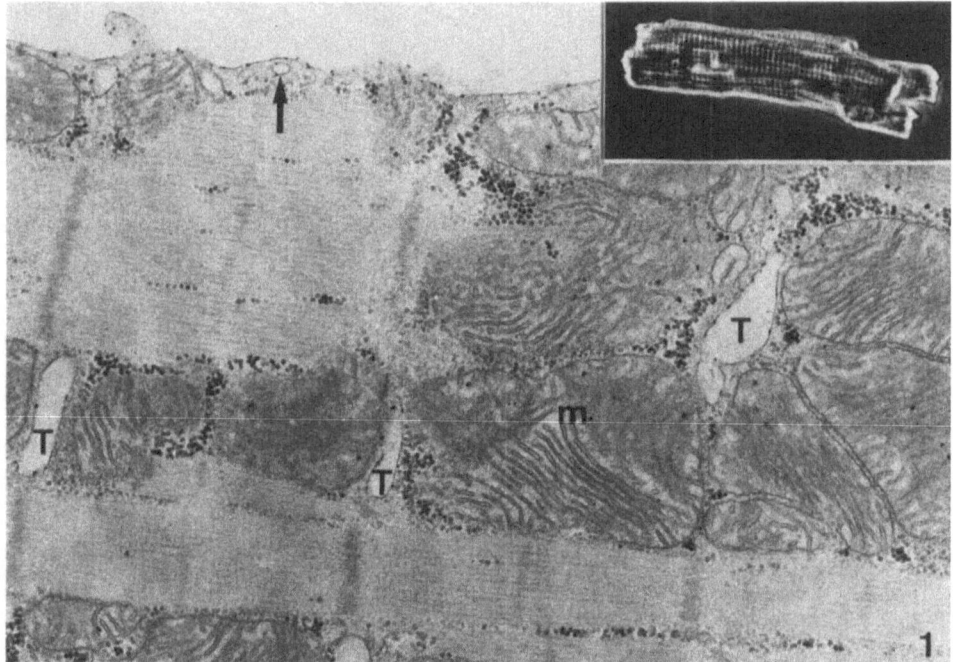

Fig. 1. Isolated Ca^{2+}-tolerant myocyte of the rat. As in the intact heart, the Ca^{2+}-deposits line the sarcolemma, sarcolemma-derived vesicles (arrows) and the swollen T-tubular membranes (T). A small number of deposits is seen in mitochondria (m). Arrowheads point to densely stained glycogen particles (X 34.560). Insert (X 200).

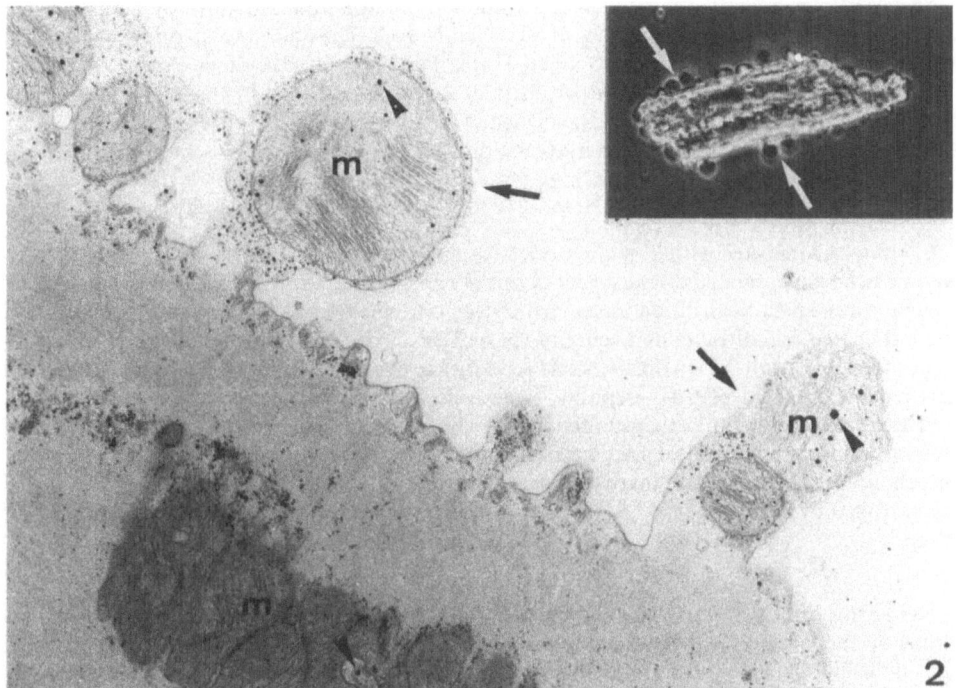

Fig. 2. Isolated myocyte exposed to 50 mM KCl for 30 min. This Ca^{2+}-intolerant cell has lost its rod-shaped appearance, formed cytoplasmic blebs (arrows) and is irreversibly contracted. The sarcolemma is devoid of Ca^{2+}-deposits and mitochondria (m) have accumulated deposits (arrowheads). (X 23.480). Insert (X 200).

that sarcolemmal Ca^{2+}-deposits persist in cells from which the external lamina of the glycocalyx is largely removed further suggest the association of Ca^{2+} with the bilayer. As in the intact heart, other organelles were practically devoid of deposits.

Ca^{2+}-Overload

The conditions which predispose the cells to Ca^{2+}-overload (Ca^{2+}-intolerance) that were chosen to assess the fate of sarcolemmal-associated Ca^{2+} were the following : a) Ca^{2+} depletion-repletion of isolated hearts (Ca^{2+}-Paradox); b) sustained depolarization of isolated myocytes in high K^+-medium; c) prolonged normothermic global ischemia of isolated hearts; d) exhaustive stimulation with high doses of epinephrine and ouabain.

Ca^{2+}-Paradox. Perfusion of isolated rat hearts with Ca^{2+}-free Krebs-Henseleit solution for 5 min induced typical changes at the cell surface whereby the external lamina of the glycocalyx becomes separated from the lipid bilayer-surface coat (11, 12). When Ca^{2+} localization was assessed in cells altered as such, the sarcolemma was totally free of Ca^{2+}-deposits. It is well established that these structural changes are the morphological expression of Ca^{2+}-intolerance, i. e. readmission of Ca^{2+} will result in an instantaneous and irreversible contracture as a consequence of intracellular Ca^{2+}-overload. This was confirmed in the present study by structural Ca^{2+}-assessment after repleting the Langendorff perfusate with normal Ca^{2+}. The sarcolemma remained devoid of Ca^{2+}-deposits whereas mitochondria accumulated large amounts of Ca^{2+}-deposits.

Sustained membrane depolarization. Ca^{2+}-intolerance was induced by incubating isolated Ca^{2+} tolerant myocytes in the presence of 50–100 mM KCl. The otherwise quiescent rod-shaped cells slowly transformed into supercontracted cells. The early phase of shape change, characterized by shortening of sarcomeres and peripheral bleb formation, was accompanied by a loss precipitable Ca^{2+} at the sarcolemma and coincided with mitochondrial Ca^{2+}-accumulation (Fig. 2). At the later phase more pronounced mitochondrial Ca^{2+} overload was present in the supercontracted cells.

Ischemia. As reported earlier in our work with dog and rabbit hearts (2, 6) complete normothermic ischemia provoked displacement of sarcolemma-bound Ca^{2+} in a substantial number of subendocardial cells. Such displacement, coinciding with disruption of the sarcolemmal integrity, was interpreted as a direct consequence of the inability of the bilayer to bind Ca^{2+} after a prolonged ischemic insult. Upon postischemic reperfusion, the sarcolemma of irreversibly damaged cells remained devoid of Ca^{2+}-deposits. Such cells also showed intracellular Ca^{2+}-overload. However, many cells of the subepicardial region which structurally recovered during reperfusion presented a normal number of sarcolemmal Ca^{2+}-deposits and a normally low load of Ca^{2+} in mitochondria. When isolated myocytes were deprived of oxygen, a loss of sarcolemmal Ca^{2+} accompanied by mitochondrial Ca^{2+}-overload was observed in a way very similar to the intact organ.

Exhaustive cardiac stimulation. The effects of a 10 min perfusion with either 10^{-4}M epinephrine or ouabain on Ca^{2+}-redistribution in rat Langendorff hearts were comparable to those after ischemia and reperfusion, i. e. loss of sarcolemma-bound deposits was succeeded or accompanied by mitochondrial Ca^{2+}-overload.

Acknowledgement

This work is supported by a grant from IWONL.

References

1. Lüllman H, Peters T (1977): Plasmalemmal calcium in cardiac excitation-contraction coupling. Clin Exp Pharm Physiol 4: 49–57
2. Borgers M, Thoné F, Xhonneux R, Flameng W, (1982): Shifts of calcium in the ischemic myocardium. A structural analysis. In: Wauquier A et al (ed). Protection of tissues against hypoxia. Elsevier Biomedical Press, Amsterdam
3. Langer GA, Frank JS, Philipson KD (1982): Ultrastructure and calcium exchange of the sarcolemma, sarcoplasmic reticulum and mitochondria of the myocardium. Pharmac Ther 16: 331–376
4. Phillipson KD, Bers DH, Hishinato AY (1980): The role of phospholipids in the Ca^{2+}-binding of isolated cardiac sarcolemma. J Mol Cell Cardiol 12: 1159–1173
5. Bianchi CP (1969): Pharmacology of excitation-contraction coupling in muscle. Fed Proc 28: 1624–1628
6. Borgers M (1983): The role of the sarcolemma-glycolcalyx complex in myocardial cell function. In: DeBakey and Gotto (eds). Factors influencing the course of myocardial ischemia. Elsevier Biomed. Press, Amsterdam.
7. Borgers M, Thoné F, Verheyen A, Ter Keurs HEDJ (1984): Localization of calcium in skeletal and cardiac muscle. Histochem J 16: 295–309
8. Borgers M, Thoné FJM, Xhonneux BJM, De Clerck FFP (1983): Localization of calcium in red blood cells. J Histochem Cytochem, 31: 1109–1116

9. Boskey AL, Posner AS (1982): Optimal conditions for Ca-acidic phospholipid-Oso$_4$ formation. Calcif Tissue Int 34: S1-S7

10. Piper HM, Probst I, Schwartz P, Hütter FJ, Spieckermann PG (1982): Culturing of calcium stable adult cardiac myocytes. J Mol Cell Cardiol 14: 397–412

11. Ruigrok TJC, Boink ABTJ, Spies F, Blok FJ, Maas AHJ, and Zimmerman ANE (1978): Energy dependence of the calcium paradox. J Mol Cell Cardiol 10: 991–1002

12. Hearse DJ, Humphrey SM and Bullock GR (1978): The oxygen paradox and the calcium paradox: two facets of the same problem? J Mol Cell Cardiol 10: 641–668

Author's address:

Dr. M. Borgers, Laboratory of Cell Biology, Department of Life Sciences, Janssen Pharmaceutica, Beerse (Belgium)

Anoxic injury of adult cardiac myocytes

H. M. Piper, P. Schwartz, R. Spahr, J. F. Hütter and P. G. Spieckermann

Zentrum Physiologie und Zentrum Anatomie der Universität Göttingen, Göttingen (F.R.G.)

Summary

Cultured adult cardiocytes were exposed to anoxia. The initial decrease of high-energy phosphates was accompanied by a moderate release of cytosolic enzymes and morphological changes: the appearance of sarcolemmal 'microblebs' (~ 1 μm in diameter) and an increase of subsarcolemmal vesicles. At ATP levels above 2 μmol/g_{ww}, metabolic and morphological alterations were reversible. Probably the sarcolemmal changes are causally related to the loss of macromolecules from reversibly injured cells. At ATP levels below 2 μmol/g_{ww}, an increasing number of cells become irreversibly hypercontracted. In these cells cytoplasmic masses are protruded into large 'macroblebs' (10–30 μm in diameter), however sarcolemmal continuity is preserved. Thus, enzyme release, irreversible contracture and cytolysis do not occur simultaneously in anoxic isolated cardiocytes.

Key words: anoxia, enzyme release, cytolysis, sarcolemmal blebbing, adult cardiac myocytes

Introduction

It has been hypothesized that irreversible cell damage, enzyme release and cytolysis are concurrent events in myocardial oxygen deficiency (1). However, because of the complexity of myocardial tissue this hypothesis cannot be proven satisfactorily in the whole heart. Isolated ventricular myocytes can be exposed to anoxia in a homogeneous environment which can be controlled immediately. In the primary culture of adult ventriculocytes used in this study (5), cells largely lie isolated from each other and therefore mechanical cell-cell interactions are absent.

Methods

A cell culture of intact, Ca^{2+}-stable ventricular muscle cells from 12-week-old Sprague-Dawley rats was prepared as previously described (5). Four hours after plating, the dishes (60 mm, Falcon) were washed and filled with 1 ml of a modified anoxic Tyrode solution (125.0 mM NaCl, 2.6 mM KCl, 1.2 mM KH_2PO_4, 1.2 mM $MgSO_4$, 1.0 mM $CaCl_2$, 25.0 mM HEPES, pH 7.4, equilibrated with 100% N_2 at 37 °C). Dishes, containing about 10^5 cells, were then transferred to an incubation chamber that was gently perfused with water-saturated nitrogen and slowly rotated (1 cycle/5 min) at a slight inclination. Re-oxygenation was performed by inflating 100%O_2 for 30 min. Under nitrogen or oxygen, supernatant was withdrawn and perchloric acid or fixative was added. Oxygen pressure was found less than 1 mm Hg after 5 min under N_2, pH was between 7.35 and 7.4 in all experiments. CP, ATP, glycogen and activities of lacate dehydrogenase (LDH, EC 1.1.1.27), malate dehydrogenase (MDH, EC 1.1.1.37) and glutamate dehydrogenase (GLDH, EC 1.4.1.3) were determined by standard enzymatic UV-methods (2). Acid phosphatase (acid P'ase, EC 3.1.3.2) activity was assayed with 4-nitrophenylphosphate as substrate, as described in (2). Enzyme activities are expressed as percent of the initial cellular content determined in cells from control samples sonicated in 1 ml Tyrode with 1% Triton X-100. Under aerobic control conditions no significant changes in high-energy phosphate contents or extracellular enzyme activities were observed during 120 min incubation. Processing of cells for electron microscopy was performed as previously described (5). By phase contrast microscopy at 400-fold magnifi-

38

cation the portion of elongated cells with discernible cross-striations vs. the portion of round cells was determined (examining more than 1200 cells at each experimental condition). If not stated differently, each datum represents the mean ± standard deviation of 5 independent experiments.

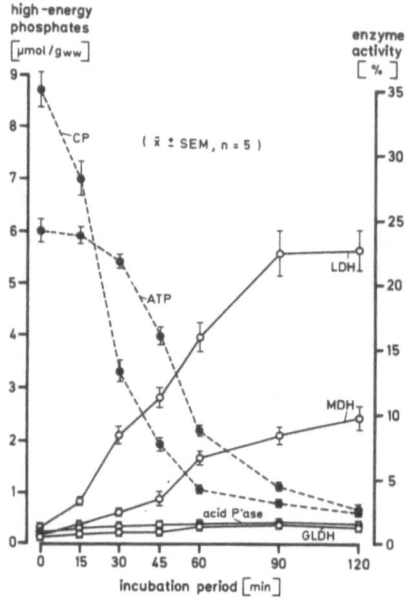

1

Fig. 1. Fall of cellular contents of CP and ATP and enzyme release (LDH, MDH, GLDH, acid phosphatase; expressed as percent of initial total activity) during anoxia. All values are given as mean ± standard error of mean from 5 independent experiments.

Results and discussion

Under aerobic control conditions, the cultured cells contain physiological levels of high-energy phosphates. Under anoxia (Fig. 1), high energy phosphate contents fall and, concomitantly, activities of cytosolic enzymes (LDH, MDH) increase in the extracellular space (15 min: $P < 0.01$, 30 min: $P < 0.001$, Kruskal-Wallis test), while lysosomal (acid P'ase) and mitochondria-specific (GLDH) enzyme activities do not. MDH, about half of which is cytosolic, is released only half as much as the cytosolic marker LDH. The anoxic release of cytosolic enzyme is highly correlated with the actual ATP content (MDH: $r = -0.98$, LDH: $r = -0.98$, for both: $P < 0.001$) and therefore slows down when ATP degradation decelerates. This deceleration is most pronounced near 2 µmol ATP/g_{ww} (60 min anoxia) where glycogenolysis also distinctly decelerates although only half of the initial amount of glycogen has been consumed (control: 43 ± 3, at 60 min: 22 ± 2, at 120 min: 18 ± 2 µmol glucose/g_{ww}). After 60 min anoxia, reoxygenation restores normal CP contents ($95 \pm 7\%$ of control), but ATP is only restored to the actual amount of total adenine nucleotides ($39 \pm 4\%$ of control).

At average ATP levels below 2 µmol/g_{ww}, more and more cells round off in hypercontracture. Initially the cell population contained $3.0 \pm 1.6\%$ round cells. Their number does not increase earlier than 90 min anoxia (60 min: $3.6 \pm 1.5\%$, 90 min: $14.1 \pm 2.5\%$, 120 min: $29.4 \pm 4.7\%$,

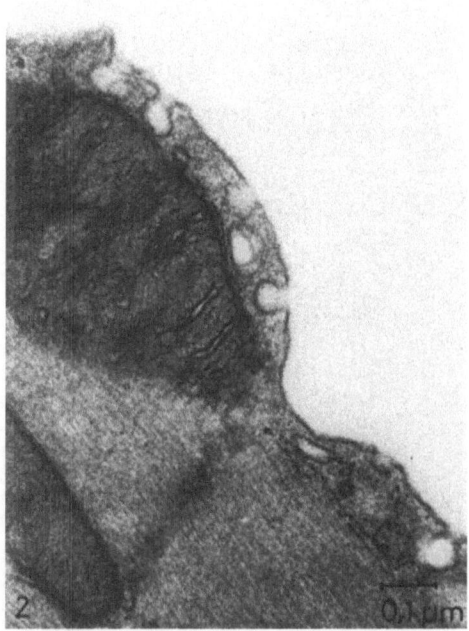

Fig. 2. Transmission electron micrograph of numerous subsarcolemmal vesicles in a myocyte at 60 min anoxia. Some of them communicate with the external space.

120 min under aerobic control conditions: $3.7 \pm 2.0\%$). Up to 120 min anoxia, reoxygenation does not change the number of round cells nor the amount of enzymes released during anoxia ($P > 0.1$, Kruskal-Wallis test) indicating that a reoxygenation damage does not occur in this system.

Thus, cytosolic enzyme release precedes irreversible cell death. Indeed it even decelerates when single cells start rounding off. As indicated by its correlation to the ATP level, early enzyme release seems to depend on availability of ATP as for an active process of exocytosis. This suggestion is further emphasized by the finding that, early in anoxia, the number of subsarcolemmal vesicles (Fig. 2) increases, and again decreases in reoxygenated cells. Since lysosomal enzymes do not increase in the external space, a lysosomal origin of these vesicles seems unlikely.

During the early phase of ATP decay, the cells gradually shorten while preserving their polygonal overall shape. Cell shortening is accompanied by the appearance of 'microblebs' (~ 1 μm in diameter), preferentially near the former intercalated disc areas. During the reversible stage of anoxia these microblebs spread over the whole surface and increase in number. At 60 min anoxia, they are found spread over the surface of all polygonal cells (Fig. 3). On reoxygenated polygonal cells microblebs are seen only occasionally. During the reversible stage of anoxia (up to 60 min) they show no tendency to become confluent. In most cases these microblebs contain one mitochondrion filling the cavity of the bleb.

In size, microblebs differ considerably from large protrusions ($10-30$ μm in diameter), typical for round cells with over-contracted, often condensed myofibrillar masses, appearing in prolonged anoxia (Fig 4). These sarcolemmal pouches contain mixtures of cell organelles or their remnants. It can be demonstrated by ruthenium-red staining that the sarcolemma with its adher-

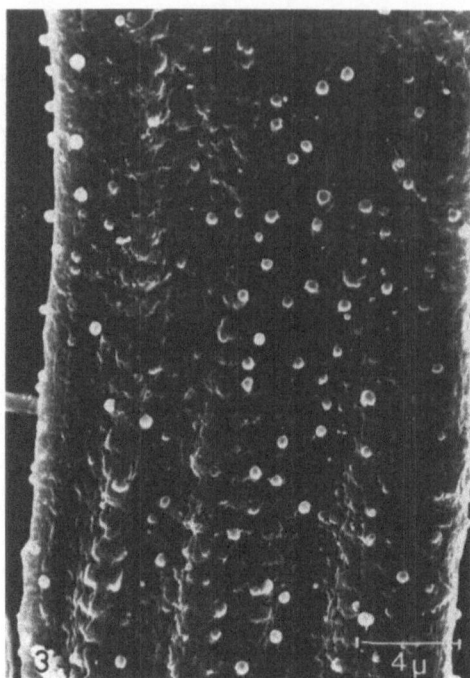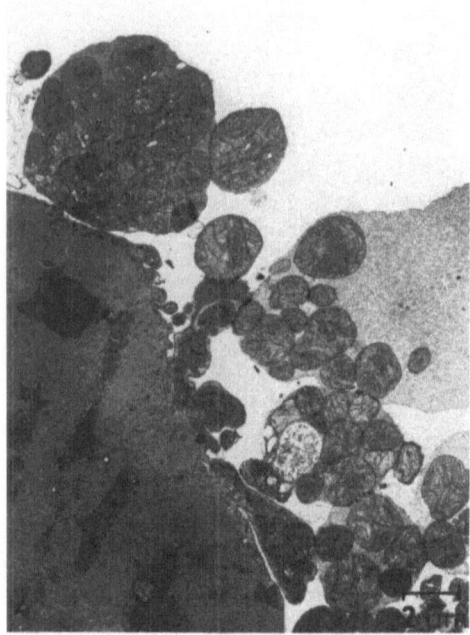

Fig. 3. Scanning electron micrograph of microblebs scattered on the lateral cell surface of a myocyte at 60 min anoxia.

Fig. 4. Transmission electron micrograph of a hypercontracted myocyte at 120 min anoxia. The surface is covered with voluminous protrusions, myofibrils are condensed to homogeneous masses.

ent surface coat is still continuous around the hypercontracted cell. This agrees with the observation that enzyme release is not accelerated when round cells appear. The ultrastructure of round cells does not differ before and after reoxygeneration. Typical features of the 'oxygen paradox' damage are absent, i. e. localized contraction bands, intramitochondrial amorphous densities and sarcolemmal discontinuities (3). It might be hypothesized that in tissue these ultrastructural changes and the massive enzyme loss, characteristic for the 'oxygen paradox', are secondary to mechanical cell-cell interactions (3) which are absent in an isolated cell system.

In conclusion, the anoxic isolated cardiocyte may already lose a moderate amount of cytosolic proteins before cell injury becomes irreversible (4). And irreversible cell injury, characterized by hypercontracture, is not necessarily accompanied by immediate cytolysis. Neither leads irreversible cell contracture to the rapid disintegration process of the 'oxygen paradox' upon reoxygenation. Early microbleb formation might be caused by local disruptions of the cytoskeleton when the intrinsic tension of the sarcolemma increases in a shortening cell. In round cells covered with macroblebs, the cytoskeletal anchoring of the sarcolemma is probably almost completely destroyed, which might also be a reason for irreversibility in this case.

Acknowledgments

This study was supported by the Deutsche Forschungsgemeinschaft, SFB 89 – Kardiologie Göttingen. For technical assistance we thank R. Zöllner, B. Eickhoff and H. Haacke.

References

1. Ahmed SA, Williamson JR, Roberts R, Clark RE, Sobel BE (1976): The association of increased plasma MB CPK activity and irreversible ischemic myocardial injury in the dog. Circulation 54: 187–193
2. Bergmeyer HU (1974): Methods of Enzymatic Analysis. Academic Press, New York
3. Ganote CE (1984): Contraction band necrosis and irreversible myocardial injury. J Moll Cell Cardiol 15: 67–73
4. Gebhard MM, Denkhaus H, Sakai K, Spieckermann PG (1977): Energy metabolism and enzyme release. J Mol Med 2: 271–283
5. Piper HM, Probst I, Schwartz P, Hütter JF, Spieckermann PG (1982): Culturing of calcium stable adult cardiac myocytes. J Mol Cell Cardiol 14: 397–412.

Author's address:

Dr. med. H. M. Piper, Zentrum Physiologie der Universität, Humboldtallee 23, D-3400 Göttingen (F.R.G.)

Non-invasive measurements of cell surface receptors

I. A. Bailey[a], V. von Tscharner[b] and D. R. Harris[a]

[a] ICI Pharmaceuticals Division, Macclesfield, Cheshire (U.K.) and [b] Theodor Kocher Institut, Universtität Bern (Switzerland)

Summary

None of the current methods for assessing the effects of subtle changes in drug structure on tissue is entirely satisfactory. The observation that ^{3}H-drug molecules emit photons of light when they bind to membrane proteins (called Condensed Phase Radioluminescence) probably represents a significant advance in measuring non-invasively kinetic effects of drugs on beating monolayers of cardiac myocytes. The potential utility of CPR is great.

Key words: condensed phase radioluminescence, drug receptor, non-invasive measurements, dihydroalprenolol, nitrendipine

Introduction

Frequently, the pharmacologist or medicinal chemist has the problem of assessing the biological effects of subtle changes in the structure of drug molecules. In heart cells in culture, this may be achieved in various ways, all of them flawed. The ultimate biological effect may be measured as, for example, beating rate enhancement in response to 5-Hydroxytryptamine (1) or as a reduction in cell death in response to calcium antagonists (2). The concentration of an intermediate or an activator may be measured, for example adenosine 3',5'-monophosphate changes in response to β-blockade (3). The effects of changes in drug structure at the receptor may be amplified or attenuated nonlinearly before their effects are measured by any of these means. The alternative is to measure drug binding directly to isolated cells (4) or, much more commonly, to membrane fragments (5). For good data to be obtained with isolated cells, one must pay careful attention to cells aggregating or sticking to apparatus, to ligands internalising or degrading, and to cell viability. Cell bursting is not uncommon in these experiments, under the stress of changes in osmolarity or temperature or during filtration. It is more convenient to use membrane fractions. Here, the problems are mainly of purity of the fraction, of receptor modification or inactivation during the membrane preparation and of presenting ligand to both sides of the membrane, unlike *in vivo*. These effects contribute to differences in potency between binding studies and *in vivo* measurements, not infrequently observed.

An entirely new approach (6) is to measure non-invasively on monolayers of cells the very low intensity light emitted when tritiated drug molecules bind to protein receptors on intact cells. The phenomenon is called Condensed Phase Radioluminescence (CPR). For theoretical reasons (7, 8), photons are only emitted in quantity when the ^{3}H-drug and the fluorophore (tryptophan in proteins) are closer together in a membrane. We have used this technique to study the kinetics of association and dissociation of a calcium antagonist and a β-blocker with isolated, neonatal cardiac myocytes in culture. The technique may prove to be a significant advance on current methods.

Methods

Suspensions of beating heart cells from neonatal rat hearts were prepared as in (3) and depleted of fibroblasts by differential adherence as in (6). The cardiac myocytes were grown on 1 cm² glass coverslips for 4 days at 37 °C, the medium (containing 10% newborn calf serum) being changed on days 2 and 4. The cells were then transported by airline cabin freight to Switzerland, where they arrived in good condition q.v. (6). All measurements were made at 20 °C on culture day 6 using an S.L.M. fluorimeter equipped for single photon counting. Coverslip cultures were added to 450 mm³ medium containing 2.1 μM ³H-nitrendipine (New England Nuclear) and the CPR signal observed until equilibrium. The medium was then changed and the kinetics of dissociation observed. The same cells continued to beat after a third medium change. Experiments were also performed on cells pre-incubated with various concentrations of nifedipine and also using 2.1 μM ³H-dihydroalprenolol (Amersham International).

Results

Fig. 1 shows that the CPR signal intensity changed on adding myocytes to medium containing ³H-nitrendipine (panel A). The background count rate in the presence of ³H-drug was very low (30 cps). When cells were added, the count rate rose in a mathematical exponential, indicating the probable presence of only one class of binding sites. The signal increase observed was in the ultraviolet frequency range, demonstrating that the binding site was a protein containing tryptophan. Adding the cells in the absence of drug caused no change in the background light, showing no interference from the cell's intrinsic chemiluminescence. Replacing the medium on the same cells with medium containing no radioactivity caused a monoexponential decrease in the CPR signal (Fig. 1C). The half-time of dissociation was 14.5 min, compared with 5.5 min for association. The cells continued to beat normally after a further medium change. Similar monoexponential

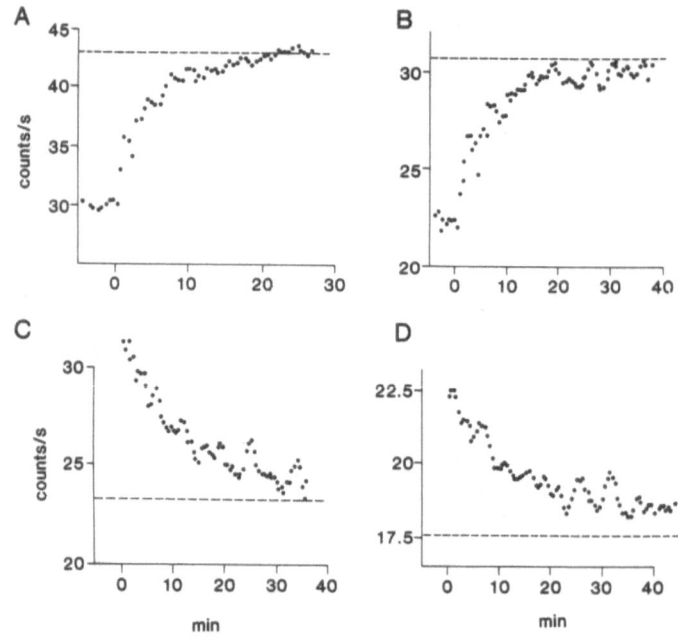

Fig. 1. The kinetics of association and dissociation of ³H-nitrendipine (panels A and C respectively) and of ³H-dihydroalprenolol (panels B and D) with cardiac myocytes in culture are shown as observed by condensed phase radioluminescence. Myocytes were added to medium containing ³H-ligand at time 0. The dotted lines represent equilibrium binding values.

kinetics of association and dissociation (Fig. 1B and C), with half-times of 6 min and 14.8 min respectively, were observed in experiments using ^{3}H-dihydroalprenolol.

Pre-incubating cells at 37 °C with 0.1 µM nifedipine before transfer to the radioactive medium reduced the amplitude of nitrendipine binding by 33% and its rate by 70%, while exerting no effect on dihydroalprenolol binding. Higher concentrations of nifedipine abolished nitrendipine but not dihydroalprenolol binding. While the precise amount of nifedipine carried over in these experiments is not known, its effect is consistent with competition with nitrendipine for the single class of proteinaceous binding sites. The absence of a nifedipine effect on dihydroalprenolol kinetics indicates that nifedipine's effect in the nitrendipine system cannot be other than at the binding site. A formal pharmacological evaluation is underway.

Discussion

The present study demonstrates that CPR is sufficiently sensitive to be used to measure the kinetics of association and dissociation of drug molecules with intact, viable cells. The monoexponential binding kinetics and the competition studies with nifedipine show that ^{3}H-nitrendipine binds to a single class of proteinaceous sites on the cell membrane, most probably the calcium slow channel. External calibration of the signal intensity would be required to relate it to the amount of material bound and hence obtain a K_D for binding.

The value of the technique lies in its non-invasive nature and in its use of only viable isolated cells (dead cells would have detached from the monolayer and not entered the sample compartment). More than one receptor can be measured sequentially on the same tissue sample, or the binding and biochemical and biological effects can be measured on the same tissue sample. There are no attendant errors associated with separating bound and free ligand as only the bound ^{3}H-drug emits photons. This may permit the measurement of kinetics at 10s time intervals. The entire process of association and dissociation can be measured on one sample within 90 min.

The technique is not simply a convenient way of measuring drug binding. Were a receptor protein to undergo a conformation change after binding ligand, then the mean distance between the tritium and the tryptophan would change. This would alter the CPR signal intensity and be visible as a multi-exponential association curve. So, it may prove possible to observe the membrane events of the transduction of binding information by CPR. Similarly, the aggregation of drug-receptor complexes may also be amenable to study. Experiments can be conducted just as easily during anoxia as in normoxia and so the anoxic alterations in receptors and receptor dynamics could be measured. As interactions between several agonists, partial agonists and antagonists of the same receptor can be studied, CPR may eventually be used to illustrate structure-activity relationships and to map receptors on living cells.

The technique of CPR still requires extensive, careful testing, but it may prove to be a very powerful aid in the study of hormone and drug effects in the future.

References

1. Higgins TJC, Bailey PJ, Allsopp D (1981): Mechanism of stimulation of cardiac myocyte beating rate by 5-Hydroxytryptamine. Life Sciences 28: 999–1005
2. Higgins TJC, Allsopp D, Bailey PJ (1980): The effect of extracellular calcium concentration and Ca-antagonist drugs on enzyme release and lactate production by anoxic heart cell cultures. J Mol Cell Cardiol 12: 909–927
3. Higgins TJC, Allsopp D, Bailey PJ (1979): The effect of β-adrenergic blocking drugs on the intrinsic beating rate of cultured myocytes. J Mol Cell Cardiol 11: 101–107
4. Lau YH, Robinson RB, Rosen MR, Bilenzitian JP (1980): Subclassification of β-adrenergic receptors in cultured rat cardiac myoblasts and fibroblasts. Circ Res 47: 41–48
5. Coleman AJ, Paterson DS, Somerville AR (1979): The β-adrenergic receptor of rat corpus luteum membranes. Biochem Pharmacol 28: 1003–1010

6. von Tscharner V and Bailey IA (1983): Non-invasive, kinetic measurements of ^{3}H-nitrendipine binding to isolated rat myocytes by condensed phase radioluminescence. FEBS Lett 162: 185–188

7. von Tscharner V and Radda GK (1980): A study of changes in surface area and molecular interactions in phospholipid residues by condensed phase radioluminescence.
Biochim Biophys Acta 601: 63–77

8. von Tscharner V and Radda GK (1981): The effect of fatty acids on the surface potential of phospholipid vesicles measured by condensed phase radioluminescence.
Biochim Biophys Acta 643: 435–448

Author's address:

I.A. Bailey, ICI Pharmaceuticals Division, Mereside, Alderley Park, Macclesfield, Cheshire SK10 4TG (U.K.)

Surface antigens of adult heart cells and their use in diagnosis*)

B. Maisch

University Hospital of Internal Medicine, Würzburg (F.R.G.)
(Director: Prof. Dr. K. Kochsiek)

Summary

The mapping of immunologically relevant epitopes of the myolemma, sarcolemma and endothelium are prerequisites for the analysis of humoral and cellular effector mechanisms. Mapping was carried out by monospecific anticollagen and anticytoskeleton antibodies, monoclonal antibodies against the surface of white blood cells and by lectins which bind to specific sugars on membranes. Characteristic differences between myolemma, sarcolemma and endothelium could be defined. – In patients with perimyocarditis and postmyocarditic cardiomyopathy antibodies directed against the myolemma were detected. They are cytolytic in the presence of complement and cross-reactive to the causative viral agents. Furthermore lymphocytotoxic effector mechanisms could be demonstrated *in vitro* in one third of patients with AMLA-negative primary dilated cardiomyopathy.

Key words: antimyolemmal antibodies, antisarcolemmal antibodies, cytotoxicity, cardiomyopathy, viral myocarditis

Introduction

The sarcolemma, like any other membrane, is composed of the structural pospholipid bilayer, the lipophilic cholesterol moieties to increase viscoelasticity, the unipolar or bipolar membrane proteins as structural elements, receptors, transport molecules and epitopes relevant for immunologic interactions, and glycoproteins, which are responsible for the unique tissue-specific and species-specific properties. Whereas the ubiquitous phospholipids are of little antigenicity the most important candidates to be recognized as antigens in an immunological reaction are the membrane proteins and the glycoproteins.

The purpose of this study was to examine the morphologic properties of the cardiac myolemma of rat myocytes and the human sarcolemma with defined monoclonal and monospecific antisera, and to assess the effect of patients' sera positive for antimyolemmal (AMLA) and antisarcolemmal (ASA) antibodies on vital cardiocytes *in vitro*. Similarly the effector mechanisms of patients' lymphocytes on isolated heart cells were examined.

Patients and methods

Isolation of adult rat cardiocytes was performed according to Powell & Twist (7). Cardiocytes were enriched to more than 80% vital cells as described previously (2,3) by Percoll gradient centrifugation. For the direct or indirect immunofluorescence test (1) cardiocytes were either used in suspension or were allowed to sediment in Sayk chambers (3). Cryostat sections from human (blood group 0) and rat heart and skeletal muscle were used similarly as previously described (3, 4). In addition in this study FITC-labelled monoclonal antibodies against surface markers of human T- and B-cells and monocytes were used (OKT 3 [=pan T]), OKT 4 (=T helper & inducer), OKT 8 (T suppressor/cytotoxic T), OKMI (monocytes), OKIA 1 (B-cells & activated T-cells); all from ORTHO diagnostic systems, Heidelberg). Preabsorbed monospecific antisera against

---

*) Supported by the Deutsche Forschungsgemeinschaft, grant MA 780/1, 2, 3.

collagen I and III were kindly provided by Drs. Bolte & Fischer (Klinikum Großhadern, München). Antilaminin and antikeratin were obtained from MEDAC (Hamburg) (FITC-labelled, dilutions 1:10 to 1:50).
Another antilaminin antibody was kindly provided by Dr. Timpl (MPI Martinsried). Sera and lymphocytes
from 38 patients with viral and 144 patients with undefined perimyocarditis (3), from 79 patients with primary
dilated cardiomyopathy, 30 patients with alcoholic and 18 patients with postmyocarditic heart disease were
analysed as described (3, 6). The FITC-labelled antibody (F[ab]$_2$-fragments) were diluted 1:50 (F/P molar ratio 2.5 ± 1.5 total protein content 25 mg/ml). The carbohydrate components of the myolemma and sarcolemma were tested with FITC-labelled lectins (MEDAC), listed in Table 2.

The pathogenic relevance of antimyolemmal (AMLA) and antisarcolemmal (ASA) as well as antiendothelial (AE) antibodies was measured in a microcytoxicity assay using vital cardiocytes as target cells (3). Antibody mediated cytolysis in the presence of complement was assessed by an index, comparing the half-life of
cardiocytes in the presence of the patient's serum and in control sera (pool). Indices smaller than 0.75 indicated significant cytolysis. In a similar assay also using vital adult rat cardiocytes as target cells direct lymphocytotoxicity (LC) and antibody-dependent cellular cytotoxicity (ADCC) were assessed (4, 5).

Results

Cross-reactivity between membrane epitopes of cardiocytes and white blood cells

Binding sites in common for white blood cells and the myolemma and sarcolemma were detected when using monoclonal OKMI- and OKT 8-antisera. No myolemmal staining on intact
cardiocytes was observed with OKT 3, OKT 10 and OKT 11, which still bound to the sarcolemma of biopsies and to the myolemma of spherical (dying) cardiocytes. This discrepancy could
either be due to newly accessible epitopes on spherical cardiocytes or to non-specific binding. The
term myolemma is used in this context only for the plasmalemma of collagenase pretreated cardiocytes deprived of perimembraneous connective tissue,the term sarcolemma for membrane and
interstitial staining in cryostat sections (thus including interstitial collagen).

Characterization of the surface antigens of the cardiocyte by monospecific antisera

Monospecific antisera against collagen I and III did not bind to the myolemma (AMLA-negative), but could be demonstrated regularly in the interstitial space between myocardial fibers in
cryostat sections (ASA-positive). Testing for cytoskeletal proteins in the myolemma gave negative results with antikeratin and antilaminin from MEDAC, whereas antilamin in antibodies
provided by the MPI in Martinsried gave a positive staining of the myolemma and the Z-bands.
This may be partly due to the differences of the species specificity of laminin used for the preparation of antilaminin antibodies.

Mapping of carbohydrate epitopes of myolemma, sarcolemma and endothelium by lectin binding

Apart from specific defense mechanisms (cytolytic T-cells and antibody-dependent effector
mechanisms) non-specific cellular and humoral factors are involved in the immune response.
Lectins represent such a biological principle, which may explain bacterial adherence, precipi-
tation of glycoproteins,tumor cell preferences, stimulation or blockade of cell receptors (8). The
results of such a mapping of cardiac and endothelial membranes by the lectins can be derived
from Table 1. The data reveal remarkable differences between rat and human myocardial membranes on one hand and myolemma, sarcolemma and vascular endothelium on the other
hand.

Antibody-dependent effector mechanisms in perimyocarditis and dilated cardiomyopathy

The characteristic immunofluorescent feature in coxsackie B, influenza A/B and mumps perimyocarditis was the presence of AMLA of the IgG-, IgM- and IgA-type in titres ranging from

Table I.

Lectin specificity	Name	Staining pattern of cardiac and endothelial membranes with:			
		intact cardioc.	spherial cardioc.	hom. myocardium	endothelium
Ph-neutral simple sugars					
Alpha-L-fucose	UEA I (Ulex europeus)	–	+	+	
L-fucose	Lotus A (+)	(+[1])	(+) –	–	+ +[2]
Beta-D-galactose	RCA I/II (Ricinus communis)	+	?	+ +	+
Alpha-D-galactose	BSA I/II, SJA (Sophora japonica)	–	–	+ +	+ +
Alpha-L-mannose	LCH. Con (Concanavalin) A. Succ. Con A	+ +	+	+ +	+ +
Hexosamines					
NAc-D-gal	DBA (Dicholus biflorus), HPA (Helix pomatia) SBA (Soy bean), VVA (Vicia villosa)	–	–	+/–	
NAc-D-glucosamin	WGA	+	+	+ +	+
Terminal di- and trisaccharides					
Beta-gal (1–3) GAL NAc	PNA (peanut agglutinin)				
alpha-D-gal (Alpha-D-fuc (1→29)–(1→3) Beta-D-gal-(1→3/4)-Beta-D-GlcNAc	EEA (Euonymus europeus)	–	–	–	+
Complex carbohydrates					
	PWM (pokeweed mitogen)	–	–	+[2]	–
	RPA (Robinia pseudoacacia)	+	+	+	+
Other lectins					
	VFA (Vicia faba)	–	+	+	+ +
N,N-Diacetylchiotobiose	STA (Solonium tuberosum)	(+)	+	–	–

An interfibrillary staining (IFA-positive) of the linear type (sarcoplasmic reticulum) was observed with lotus A, MPA, LCH, Con A, Succ-Con A, PHA, CSA, WGA, STA & VFA.

A Z-Band pattern – sometimes indistinguishable from the actin-binding site – was observed with BSA I/II, SBA & RPA.

An anti-actin pattern was observed with PHA-L.

[1] dotted granular form

[2] with cryostat sections pretreated with crude collagenase for 15 minutes the positive staining was abolished

1:40 to 1:320 (3). Examination of cryostat sections revealed that heart muscle specific AMLA were associated in more than 50% of cases with non-organ-specific antiendothelial antibodies (AEA). In postmyocarditic cardiomyopathy ASA were demonstrated in 94% of cases (6). This pattern differed significantly ($p<0.0001$) from that of patients with alcoholic or primary dilated cardiomyopathy (AMLA 9% positive, ASA 10% positive). ASA were not only found circulat-

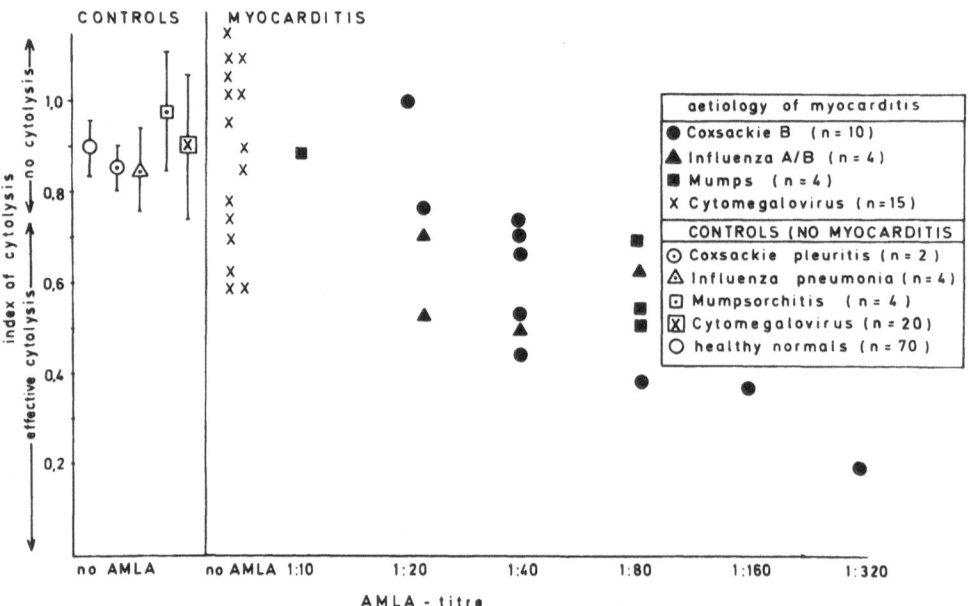

Fig. 1. Cytolytic serum activity in viral perimyocarditis correlated with the titer of antimyolemmal antibodies.

Table 2. Antibody independent lymphocytotoxicity (LC) and blocking factors in primary and secondary cardiomyopathies

Type of dilated cardiomyopathy	Lymphocytotoxicity (K-cell-activity)		Blocking factors	
	CI_{LC}*) ($\bar{x} \pm SD$)	incidence positive***) (%)	CI_{LBF}**) ($\bar{x} \pm SD$	incidence of blocking factors present***) (%)
primary (n = 36)	0.72 ± 0.14	33	0.77 ± 0.3	8
secondary (n = 25)				
-alcoholic (n = 14)	0.82 ± 0.27	29	1.01 ± 0.16	22
-postmyocarditic (n = 11)	0.89 ± 0.16	9	1.01 ± 0.13	9

*) cardiocyte index of lymphocytotoxicity
**) cardiocyte index of lymphocytotoxicity blocking factors
***) index < 0.75; $CI_{LC} = \dfrac{\text{half life of cardiocytes in presence of pat. lymphocytes after 24 h}}{\text{half life of cardiocytes in presence of control lymphocytes after 24 h}}$
The CI_{LCBF} is defined accordingly. In this assay the patient's serum was added.

ing in the patients' sera but also bound to the autologous sarcolemma in the left ventricular endo-myocardial biopsy specimen (66% positive in postmyocarditic heart disease). In acute perimyo-carditis AMLA-positive bera induced cytolysis of vital cardiocytes in the presence of comple-ment. The degree of cardiocytolysis as expressed by the CI_{AMC} correlated with the titer of AMLA (Fig. 1). This cardiocytolytic effect could be abolished if complement was omitted. It could be di-minished substantially by preabsorption of the sera with the respective causative viral agent. This indicates that cross-reactive cytolytic AMLA are operative in viral perimyocarditis, which points to a secondary immunopathogenesis in these patients. Similarly only in postmyocarditic, AM-LA-positive cardiomyopathy was cytolytic serum activity observed (6).

Whereas in acute perimyocarditis and postmyocarditic cardiomyopathy no significant lymphocyte-mediated cytotoxicity against cardiac myocytes *in vitro* was demonstrated, 33% of patients with primary and 29% of patients with secondary alcoholic cardiac disease showed im-pressive lymphocytotoxicity. After addition of autologous serum the incidence of cardiocytotox-icity decreased only slightly to 25% in primary cardiomyopathy, whereas in alcoholic heart dis-ease it was virtually abolished (Table 2).

Conclusions

The mapping of characteristic moieties of the myolemma, sarcolemma and endothelium are prerequisites for the definition of heart muscle-specific epitopes which are targets of humoral and cellular effector mechanisms. In perimyocarditis and in most patients with postmyocarditic car-diomyopathy cytolytic crossreactive AMLA are operative. In 30% of patients with primary di-lated cardiomyopathy different lymphocytotoxic effector mechanisms *in vitro* were detected. Im-munological data thus lead to a new classification of cardiomyopathies.

References

1. Coons AH, Kaplan MM (1950): Localization of antigens in tissue cells. I. Improvements in a method for the detection of antigen by means of fluorescent antibody. J Exp Med 91: 1–13
2. Maisch B (1981): Enrichment of vital adult cardiac muscle cells by continuous silica sol gradient centri-fugation. Basic Res Cardiol 76: 622–629
3. Maisch B, Trostel-Soeder R, Stechemesser E, Berg PA, Kochsiek K (1982): Diagnostic relevance of hu-moral and cell-mediated immune reactions in patients with acute viral myocarditis. Clin Exp Immu-nol 48: 533–545
4. Maisch B, Eichstaedt H, Kochsiek K (1983): Immune reactions in infective endocarditis I. Clinical data and diagnostic relevance of antimyocardial antibodies. Am Heart J 106: 329–337
5. Maisch B, Mayer E, Schubert U, Berg PA, Kochsiek K (1983): Immune reactions in infective endocardi-tis. II. Relevance of circulating immune complexes,serum inhibition factors, lymphocytotoxic reactions, and antibody-dependent cellular cytotoxicity against cardiac target cells. Am Heart J 106: 338–344
6. Maisch B, Deeg P, Liebau G, Kochsiek K (1983): Diagnostic relevance of humoral and cytotoxic im-mune reactions in primary and secondary dilated cardiomyopathy. Am J Cardiol 52:1072–1078
7. Powell R, Twist VW (1976): A rapid technique for the isolation and purification of adult cardiac muscle cells having respiratory control and a tolerance to calcium. Biochem Biophys Res Commun 72: 327–333
8. Uhlenbruck GG (1983): Die Biologie der Lektine: Eine biologische Lektion. Funkt Biol Med 2: 40–48

Author's address:

Prof. Dr. Bernhard Maisch, Medizinische Universitätsklinik, Josef-Schneider-Str. 2, D-8700 Würzburg (F.R.G.)

Substrate utilization of adult cardiac myocytes

R. Spahr, I. Probst, and H. M. Piper

Zentrum Physiologie und Zentrum Biochemie der Universität Göttingen, Göttingen (F.R.G.)

Summary

Cultured adult myocytes are in a state of basal metabolism. When glucose is the only exogenous substrate, they produce lactate over CO_2 at a constant rate of 2.7 from this substrate. Increase of oxygen tension does not change this behaviour. Insulin preferentially increases lactate formation, dichloroacetate only CO_2 production. The fact that the lactate/CO_2 ratio can be varied from 0.5 to 16 indicates that there is no close coupling between glycolytic flux and pyruvate oxidation. Both exogenous lactate and fatty acids are used preferentially over glucose. But increase of fatty acid oxidation and inhibition of glucose oxidation are not complementary. Glycolytic flux is only slightly decreased when fatty acid oxidation is already saturated. The results indicate that fatty acids interact with glucose oxidation primarily by inactivation of the pyruvate dehydrogenase. Neither insulin nor dichloroacetate in the presence of glucose inhibit fatty acid oxidation.

Key words: carbohydrates, palmitate, insulin, dichloroacetate, adult cardiac myocytes

Introduction

At physiological Ca^{2+} concentrations isolated ventricular muscle cells are mechanically at rest. Thus, with the isolated cell system one might study the basal metabolic activity of the myocardium in the absence of contractile performance. This may allow to investigate in more detail and with better resolution those metabolic processes whose energy needs are very small compared to the energy needed for contraction. In contrast to the arrested heart, in the isolated cell system this can be done in physiological media.

In this study we have investigated some features of the interaction of the utilization of carbohydrates and fatty acids in the myocardial cell. Earlier studies of the metabolic behaviour of isolated cardiocytes were partly performed with preparations that were not Ca^{2+}-tolerant, and mostly shaken or stirred suspensions were used (1, 5, 6, 14). However, such mechanical agitation may increase the cellular energy turnover, probably due to the stimulation of contractions (8). Therefore, it is difficult to relate the results of such studies to a defined physiological state. Additionally, the cell material used in those studies contained 10 to 40% severely damaged cells. Their contribution to the observed effects is difficult to estimate. Our investigations were performed with a primary culture of adult cardiocytes which contained more than 95% intact cells that are mechanically at rest (8).

Methods

Ventricular muscle cells were isolated from 12 weeks old female Sprague-Dawley rats and plated on 60 mm dishes in M-199 medium containing 4% fetal calf serum, as previously described (8). After 4 hours of incubation, dishes were washed and filled with 3 ml of a modified Tyrode solution (basic composition: 125.0 mM NaCl, 2.6 mM KCl, 1.2 mM KH_2PO_4, 1.2 mM $MgSO_4$, 1.0 mM $CaCl_2$, 10.0 mM HEPES, pH 7.4, 37 °C), supplemented with various substrates. Each dish was then transferred to an air-tight incubation chamber filled with air and incubated at 37 °C. As substrates D-glucose, L-lactate and palmitic acid were used. As previously described (9), palmitic acid was bound to bovine serum albumin ('essentially fatty acid free' from Sigma, Tauf-kirchen, FRG) in various ratios. Albumin was tested not to change glucose utilization. Culture dishes were provided with medium containing substrates and insulin as indicated, including U-^{14}C substrates (0.3–0.6 μCi per plate). After 45 min incubation at 37 °C, single dishes were taken as zero-time samples. The experiments were terminated 2 hours later by injection 0.5 ml 1 N HCl through the stoppers of the incubation chamber into the medium. The generated CO_2 was quantitatively collected in an inbuilt reservoir of 0.5 ml 2 N KOH. Radioactive glucose and lactate were separated by ion-exchange chromatography (10). The radio-active fractions were mixed with Supersolve scintillator (Zinsser, Frankfurt, FRG) and counted directly. All values were expressed as referring to wet weight units. This was performed by relating data first to ATP contents of control dishes taken before the start which were shown to contain 6.0 ± 0.2 μmol ATP/g_{ww} in a large series of preparations ($\bar{x} \pm$ SD, n = 120). If not stated differently, results are presented as mean ± standard error of mean of 5 independent experiments.

Results and discussion

When glucose (5 mM) is the only external oxidizable substrate, the cells produce lactate and CO_2 linearly for hours at a constant molar ratio of 2.7 ± 0.5. This means that only one out of ten glucose molecules degraded is completely oxidized to CO_2. This ratio is not changed under 100% oxygen, thus the high lactate output is not due to hypoxic conditions. Isolated hearts can also produce lactate when glucose is the sole external carbon source and the reason for this is believed not to be cellular hypoxia (3, 13). However, in the beating heart CO_2 production largely exceeds lactate formation. This discrepancy might be partially explained by a less active pyruvate dehydrogenase (PDH) in the resting cell compared to the beating organ. Indeed in the presence of dichloroacetate (DCA) which inhibits PDH phosphorylation (12), flux through pyruvate dehydrogenase can be significantly increased up to 4 fold with 10 mM DCA whereas lactate production is unaffected. The lactate/CO_2 ratio therefore decreases below 1 (Fig. 1). The fact that pyruvate oxidation can be stimulated shows that the activity of the mitochondrial pyruvate carrier is not limiting and thus not responsible for the high lactate output.

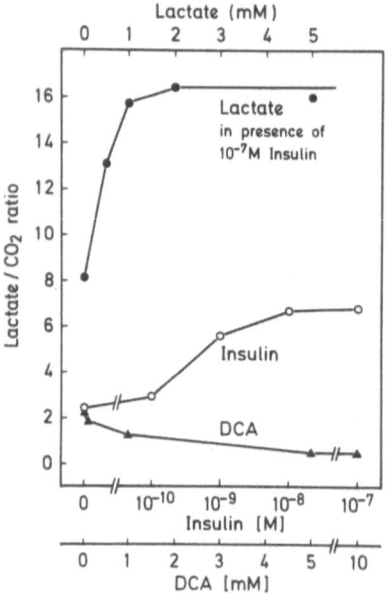

Fig. 1. Ratio of lactate to CO_2 produced from exogenous glucose (5 mM) under increasing medium concentrations of L-lactate (in presence of 10^{-7} M insulin, closed circles), of insulin (open circles) and of dichloroacetate (DCA, triangles). Mean values of 5 independent experiments.

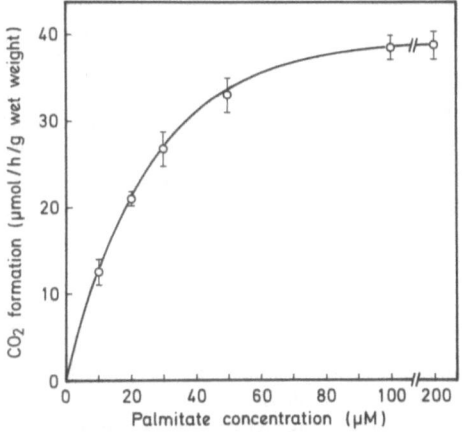

Fig. 2. CO_2 production from palmitate bound $5:1$ to albumin (in the presence of 5 mM glucose; $\bar{x} \pm SEM$, $n = 5$).

Whereas variation of glucose from 2 to 10 mM does not change the rate of glucose degradation, insulin at saturating concentration (10^{-7} M) leads to a 4.5 ± 0.2 fold stimulation of lactate production and stimulates CO_2 formation by $56 \pm 5\%$. This again increases the lactate/CO_2 ratio up to 7 fold (Fig. 1). Half maximal effective dose of insulin is 10^{-9} M which agrees with data reported for the stimulation of the glucose transporter (4). Thus we can assume that the increased glycolytic flux is caused by insulin-stimulated glucose entry into the cell and presumably by subsequent activation of phosphofructokinase (7). However, the increase in glycolytic flux preferentially leads to enhanced lactate formation, as this seems to be the main route by which cytosolic NADH can be reoxidized. In comparison to the beating organ the stimulating effect of insulin is much bigger in this *in vitro* system. It is probable that mechanical activity by itself leads to increase in glucose uptake and glycolysis (7).

In vivo there is no situation in which glucose is the only carbon and energy source, in fact lactate and fatty acids are preferentially oxidized (11, 15). In cultured cells, glucose oxidation to CO_2 is more sensitive to low lactate concentration than glucose conversion to lactate, which again leads to a change in the lactate/CO_2 ratio. This ratio, measured in the presence of 10^{-7} M insulin, could be as high as 16 at 5 mM lactate (Fig. 1). The variability of the lactate/CO_2 ratio, as shown in Fig. 1, illustrates that in this cell system there is no close coupling between glycolytic flux and the oxidative degradation of pyruvate.

Oxidation of palmitate is saturated at 100 µM palmitate (Fig. 2). However, conversion of glucose to lactate is unchanged and conversion to CO_2 is only partially (31%) inhibited. The latter further decreases linearly with increasing palmitate concentration (at 50 µM palmitate $86 \pm 6\%$ of control, at 100 µM $69 \pm 5\%$, at 200 µM $43 \pm 3\%$). Palmitate-induced inhibition of glucose-oxidation by 30% decreases glycolytic flux by only 3%. DCA is able to completely relieve palmitate inhibition of CO_2 production. It seems therefore, that palmitate exclusively inhibits glucose oxidation through decrease in pyruvate dehydrogenase activity. When the glucose-medium is supplemented with palmitate, saturating fatty acid concentrations produce 4 times as much CO_2 per hour as glucose does. At present it is not yet known if this increase also holds for oxygen consumption. However it is improbable that this increase is due to an uncoupling effect of fatty acids since then glucose oxidation should also be stimulated.

No change in the palmitate substrate saturation curve is detected when the palmitate/albumin ratio is varied between $1:1$ and $5:1$. This finding contrasts with the demonstration that in the working heart palmitate oxidation is dependent on the albumin-bound palmitate fraction. This latter finding led to the postulation that an albumin mediated uptake is the limiting step in palmitate oxidation (2). However, in the cell system a later step, for instance acetyl-CoA oxidation, might be limiting (7) and the specific transport mechanism in the plasma membrane might work far below its maximal capacity due to the much lower palmitate oxidation in this system. Thus there may be enough time for complete equilibration of palmitate between the medium and the sarcoplasm through both a receptor-mediated uptake from the albumin-bound fraction and an uptake by diffusion from the unbound fraction. Palmitate oxidation could not be lowered by the addition of insulin or DCA, two substances which accelerate glucose degradation. This again questions the proposal that a switch from fatty acid to carbohydrate oxidation, which might help to save oxygen, could be achieved by administration of insulin and glucose.

References

1. Burns AH, Reddy WJ (1978): Amino acid stimulation of oxygen and substrate utilization by cardiac myocytes. Am J Physiol 234: E461–E466
2. Hütter JF, Piper HM, Spieckermann PG (1984): Kinetic analysis of myocardial fatty acid oxidation suggesting an albumin receptor mediated uptake process. J Mol Cell Cardiol 16: 219–226
3. Kobayashi K, Neely JR (1979): Control of maximum rate of glycolysis in rat cardiac muscle. Circ Res 44: 166–175
4. Lindgren CA, Paulson DJ, Shanahan MF (1982): Isolated cardiac myocytes. A new cellular model for studying insulin modulation of monosaccharide transport. Biochim Biophys Acta 721: 385–393
5. Liu MS, Spitzer JJ (1978): Oxidation of palmitate and lactate by beating myocytes isolated from adult dog heart. J Mol Cell Cardiol 10: 415–426
6. Montini J, Bagby GJ, Spitzer JJ (1981): Importance of exogenous substrates for the energy production of adult rat heart myocytes. J Mol Cell Cardiol 13: 903–911
7. Morgan HE, Neely JR (1982): Metabolic regulation and myocardial function. In: Hurst JW (ed), The Heart, pp. 128–142. Mc Graw Hill, New York
8. Piper HM, Probst I, Schwartz P, Hütter JF, Spieckermann PG (1982): Culturing of calcium stable adult cardiac myocytes. J Mol Cell Cardiol 14: 397–412
9. Piper HM, Sezer O, Schwartz P, Hütter JF, Spieckermann PG (1983): Fatty acid-membrane interactions in isolated cardiac mitochondria and erythrocytes. Biochim Biophys Acta 732: 193–203
10. Probst I, Schwartz P, Jungermann K (1982): Induction in primary cultures of 'gluconeogenic' and 'glycolytic' hepatocytes resembling periportal and perivenous cells. Eur J Biochem 126: 271–278
11. Randle PJ, Newsholme EA, Garland PB (1964): Regulation of glucose uptake by muscle. 8. Effects of fatty acids, ketone bodies and pyruvate, and of alloxan-diabetes and starvation, on the uptake and metabolic fate of glucose in rat heart and diaphragm muscles. Biochem J 93: 652–665
12. Randle PJ (1976): Regulation of glycolysis and pyruvate oxidation in cardiac muscle. Circ Res 38, Suppl I: 8–15
13. Strong P, Mullings R, Illingworth JA (1979): Aerobic lactate synthesis by cardiac muscle. Eur J Biochem 102: 625–636
14. Vahouny GV, Wei RW, Tamboli A, Albert EN (1979): Adult canine myocytes: isolation, morphology and biochemical characteristics. J Mol Cell Cardiol 11: 339–357
15. Williamson JR (1962): Effects of insulin and diet on the metabolism of L (+)-lactate and glucose by the perfused rat heart. Biochem J 83: 377–383

Authors' address:

R. Spahr, Zentrum Physiologie, Universität Göttingen, Humboldtallee 23, D-4300 Göttingen (F.R.G.)

Characteristics of lactate transfer in isolated cardiac myocytes

H. Kammermeier, B. Wein, and W. Graf

Dept. Physiology, Medical Faculty, RWTH Aachen, Aachen (F.R.G.)

Summary

L-lactate uptake of isolated cardiac myocytes was investigated, since due to different lactate concentrations in the interstitial fluid and vascular space, lactate uptake cannot be studied satisfactorily in whole hearts. Lactate uptake exhibits sigmoidal saturation kinetics. Pyruvate (2.3 mM) inhibits L-lactate uptake at lower lactate concentrations (< 15 mM) and enhances L-lactate uptake at higher (> 25 mM) lactate concentration. L-lactate uptake is increased at lowered pH (7.1) to an extent not explainable by non-ionic diffusion. The results are discussed in terms of a complex L-lactate carrier system which might involve cooperative mechanisms and H^+–co– or OH^--countertransport.

Key words: lactate, carrier, pyruvate, pH-dependence, sarcolemma

Introduction

Lactate besides glucose and UFA, particularly under increased work load of the total body, is a predominant substrate for cardiac energy metabolism. Whereas glucose transfer mechanisms, of the whole heart at least, are thoroughly investigated (1, 9, 12) lactate transfer has not yet been studied in detail. As demonstrated in a previous publication investigation of uptake processes in the whole heart is only partly representative for uptake mechanisms of the cardiac myocytes (4). This is a consequence of the distinct barrier functions of the capillary wall, which has its specific transport mechanism for glucose (6) and leads to interstitial substrate concentrations which differ markedly from those of plasma (15). This is true, particularly for lactate, the uptake process of which has not yet been studied independently of capillary transfer in isolated myocytes to the best of our knowledge.

Methods

For isolation of cardiac myocytes from female SPF-Sprague Dawley rats (180–220 g), the hearts were perfused by the Langendorff-technique with a collagenase containing medium in the same way and using the same media published previously (4). In most recent experiments a modified medium with a calcium content above 30 µM was used, keeping hearts and cells at a temperature not lower than 22 °C, leading to calcium-resistant cells. The results of these experiments did not differ from those which were obtained by the method published previously. Also tracer incubation, separation by centrifugation through a silicon oil layer and estimation of the initial uptake rate by Gauss-Newton iteration was performed as published (4).

Results

The uptake rate of L-lactate in glucose (5.5 mM)-containing medium depends on the extracellular L-lactate concentration (Fig. 1) and reaches saturation at an extracellular concentration of about 12 mM. The shape of the curve apparently is not of the Michaelis-Menten type but rather sigmoidal. Accordingly, a double reciprocal plot does not yield a straight line. With pK of lactate 3.86 non-ionic diffusion of lactic acid plays a role in other cell types (8, 14). Taking data of literature (2) for permeability for non-ionic diffusion of lactic acid through red cell membranes and using an estimated volume surface area ratio of $2.56 \cdot 10^{-4}$ cm (11) the non-ionic transfer of lactic acid through the sarcolemma was calculated. It amounts to less than 10% of the total uptake estimated (Fig. 1) in the concentration range of lactate of about 3.5 up to 60 mM. Since many carrier systems of monocarboxylic acids are not highly specific for one substrate and since pyruvate is used as additional substrate for isolated heart preparations frequently, the influence of pyruvate on lactate uptake was investigated in addition. Under the influence of pyruvate (2.3 mM) lactate uptake is markedly changed in a very different way at low (<15 mM) as compared to high (>25 mM) lactate concentration (Fig. 1). In the lower concentration range lactate uptake is depressed by pyruvate down to 30 to 40%. At higher concentrations L-lactate uptake is enhanced up to 200 to 800%.

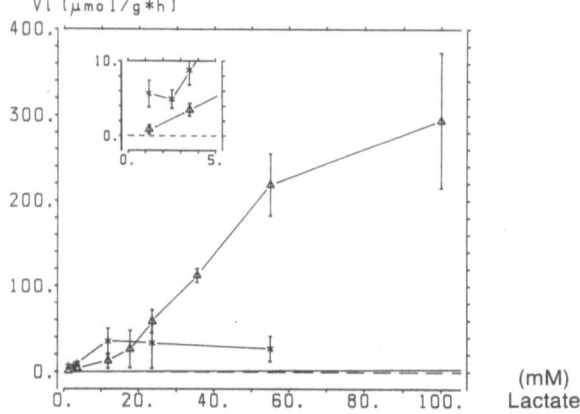

Fig. 1. L-lactate uptake of isolated cardiac myocytes (x) as influenced by 2.3 mM pyruvate (Δ). The insert shows an expended scale graph of the lower left part of the graph. The straight line just obove the dotted zero-uptake line represents the calculated non-ionic transfer.
$\bar{x}$, $\pm$ SEM, n: 3–8.
The differences are statistically significant (p <0.05, t-test) at lactate concentrations of 1.2, 3.5 and 55.0 mM.

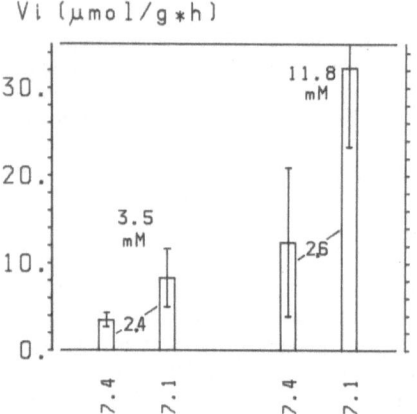

Fig. 2. Influence of pH on lactate uptake of isolated cardiac myocytes at two extracellular lactate concentrations: $\bar{x}$, $\pm$ SEM, n: 3–6.

In other cell types lactate transfer by monocarboxylic acid transporters is reported to depend also on extracellular pH (8,14). This could also be demonstrated in our study. Lowering pH from 7.4 to 7.1 i. e. doubling H^+-ion concentration, the uptake rate of L-lactate in isolated cardiomyocytes increases by a factor of about 2.5 at both concentrations 3.5 and 11.8 mM (Fig. 2).

Discussion

The results presented yield three essential findings: i. e. first the type of kinetics of L-lactate uptake, second its dependence on pyruvate and third its dependence on pH.

The uptake kinetics with saturation and sigmoidal shape indicates carrier mediated uptake, however, not of simple substrate-carrier interaction determined by mass action ratio, since the latter should exhibit Michaelis-Menten-like kinetics. It rather points to a cooperative carrier system. When formally calculating according to Segel (13) a cooperative system with 4 binding sites would fit best to the data points. However, data of literature on cooperative carrier systems for lactate transfer are not available.

The second finding of the influence of pyruvate on L-lactate uptake with different characteristics at high and low L-lactate concentrations indicates mediated transport as well. The inhibitory effect at low L-lactate concentrations is explained best by competition of lactate and pyruvate for the carrier binding site. Examples of competitive interaction of pyruvate and lactate are described for other systems in literature (5, 8, 10, 14). However, for a definite statement systematic variations of pyruvate and lactate concentrations are necessary, which are now under study.

The third finding of pH-dependence of L-lactate transfer has to be considered from two aspects. Lowering pH from 7.4 to 7.1 increases the H^+-ion concentration and the concentration of non-ionic lactic acid by a factor of 2. However, since non-ionic transport can be estimated to contribute only less than 10% to the total transport at pH 7.4, doubling of this fraction of transport cannot be responsible for an increase by a factor of about 2.5 in the total transport. Thus, the more probable explanation for pH-dependence is the existence of a H^+-co- or OH^--countertransport. This type of coupled transport is reported by various authors for other systems (8, 14), where the discrimination between both is not yet possible.

All three findings point to a rather complex carrier mediated transport system for L-lactate in the sarcolemma of cardiac myocytes. Detailed investigation, however, as for instance of the glucose transfer system (4) is hampered by the lack of suitable analogues of lactate.

Finally, quantitative aspects of L-lactate uptake are to be considered: according to our results L-lactate uptake at high but physiological concentrations of 11.8 mM accounts to $34 \mu M \cdot g^{-1} \cdot h^{-1}$, which is equivalent to an oxygen consumption of $42 \mu l \cdot g^{-1} \cdot min^{-1}$. Under physiological conditions this high lactate concentration is accompanied by severe work load, increased sympathetic stimulation and cardiac oxygen consumption of $200-400 \mu l \cdot g^{-1} \cdot min^{-1}$. Under these circumstances lactate serves as a predominant substrate of cardiac energy metabolism (7). This figure seems to be in contrast with our findings, however, in our experimental set-up the myocytes are not beating and oxygen consumption is as low as about $50 \mu l \cdot g^{-1} \cdot min^{-1}$, a value similar to that published by other authors (3). Thus, if lactate uptake does not only depend on lactate concentration but also on work load, metabolic rate or oxygen consumption, our data and those observed in whole animals are compatible.

References

1. Cheung YJ, Conover Ch, Regen DM, Whitefield CF, Morgan HE (1978): Effect of insulin on kinetics of sugar transport in heart muscle. Am J Physiol 234: E70–E78
2. Deuticke B, Beyer E, Forst B, (1982): Discrimination of three parallel pathways of lactate transport in the human erythrocyte membrane by inhibitors and kinetic properties. Biochim Biophys Acta 684: 96–110

3. Farmer BB, Harris RA, Jolly WW, Hathaway DR, Katzberg A, Watanabe AM, Withlow AL, Besch HR (1977): Isolation and characterization of adult rat heart cells. Arch Biochem Biophys 179: 545–558
4. Gerards P, Graf W, Kammermeier H (1982): Glucose transfer studies in isolated cardiocytes of adult rats. J Mol Cell Cardiol 14: 141–149
5. Halestrap AP (1976): Transport of pyruvate and lactate into human erythrocytes. Biochem J 156: 193–207
6. Kammermeier H, Kasenda S, Jung T, Lang U, Graf W (1979): Myocardial transendothelial uptake of substrates and metabolites during dual tracer bolus administration. Microvasc Res 17: S77
7. Keul J, Doll E, Stein H, Fleer U, Reindell H (1965): Über den Stoffwechsel des menschlichen Herzens. III. Der oxidative Stoffwechsel des menschlichen Herzens unter verschiedenen Arbeitsbedingungen. Pflügers Arch 282: 43–53
8. Moll W, Girard H, Gros G (1980): Facilitated diffusion of lactic acid in the guinea-pig placenta. Pflügers Arch 385: 229–238
9. Morgan HE, Regen DM, Park CR (1964): Identification of a mobile carrier mediated sugar transport system in muscle. J Biol Chem 239: 369–374
10. Oldendorf WH (1973): Carrier mediated blood-brain transport of short-chain monocarboxylic organic acids. Am J Physiol 224,6: 1450–1453
11. Page E, McCallister LP (1973): Quantitative electron-microscopic description of heart muscle cells. Am J Cardiol 31: 172–181
12. Post RL, Morgan HE, Park CR (1961): Regulation of glucose uptake in muscle. III. The interaction of membrane transport and phosphorylation in the control of glucose uptake. J Biol Chem 236: 269–272
13. Segel IH (1975): Adair-Pauling simple sequential interaction model. In: Segel IH (ed), Enzyme Kinetics. J Wiley & Sons Inc, New York/London/Sydney/Toronto
14. Spencer TL, Lehninger AL (1976): L-lactate transport in Ehrlich ascites tumor cells. Biochem J 154: 405–414
15. Wendtland B, Jüngling E, Kammermeier H (1982): Interstitial fluid concentration of metabolites in isolated perfused rat hearts as function of myocardial and capillary permeability. Pflügers Arch 394: R 11

Author's address:

H. Kammermeier, Dept. Physiology, Medical faculty, RWTH Aachen, Pauwelsstraße, D-5100 Aachen (F.R.G.)

The insulin receptor of adult heart muscle cells

J. Eckel and H. Reinauer

Diabetes Research Institute, Biochemical Department, Düsseldorf (F.R.G.)

Summary

Isolated cardiac myocytes possess specific receptors for insulin which consist of high- and low-affinity sites and are randomly distributed at the cell surface at 37 °C. Low-affinity sites can mediate a biological response comparable to that of high-affinity sites. Calcium appears to modulate the high-affinity site and thus may be involved in the regulation of cardiac sensitivity towards insulin. The data suggest involvement of the insulin receptor in insulin degradation, which includes internalization and lysosomal processing of insulin.

Key words: insulin, insulin receptor, insulin degradation, calcium, hormone internalization

Introduction

The first step in the action mechanism of insulin is binding to specific receptor sites on the surface of target cells (1). The kinetics of this binding reaction has been extensively studied in a variety of cell and membrane preparations from different tissues (2–4). Comparable information concerning cardiac muscle is presently lacking, though the heart is known to be an important target of insulin (5).

Calcium-stable adult cardiac myocytes have been recently used by our laboratory in order to study insulin binding (6–10) and to correlate it to insulin action (11–13) in this tissuse. In the present paper we report on the basic characteristics and functional aspects of cardiac insulin receptors. Moreover, data concerning the receptor-mediated processing of insulin by isolated cardiac myocytes will be presented.

Methods

Calcium-tolerant myocytes from adult rat were isolated by perfusion of the heart with collagenase followed by additional trypsin treatment, as previously described (11). Measurements of insulin binding and 3-O-methylglucose transport were performed in Hepes buffer and quantified by the oil centrifugation technique (7, 11). Gel filtration was performed on a sephadex G-50 column as detailed elsewhere (9).

Results and discussion

Basic characteristics and functional aspects of cardiac insulin receptors

Incubation of cardiac myocytes with ^{125}I-labelled insulin at 37 °C resulted in a rapid association of the hormone reaching maximum binding by 60 min. Insulin binding increases at lower temperatures, as observed for human monocytes (14), rat liver (15) and kidney plasma membranes (16). It is noteworthy that insulin degradation never exceeded 2–3% when cardiac cells were incubated in the presence of albumin (2%). Peptide hormones unrelated to insulin (ACTH, TSH, glucagon) did not affect the binding reaction, demonstrating the specificity of cardiac insulin receptors. Under steady-state conditions (60–180 min, 37 °C) all the receptors appeared to be maintained in equilibrium with extracellular insulin, since 90% of specifically bound hormone could be dissociated from the cells by addition of excess insulin (1 μM). These data contrast with a report by Kahn and Baird (17) on isolated adipocytes, which suggested that the hormone is rapidly transferred into a 'compartment' which is less accessible from the external milieu.

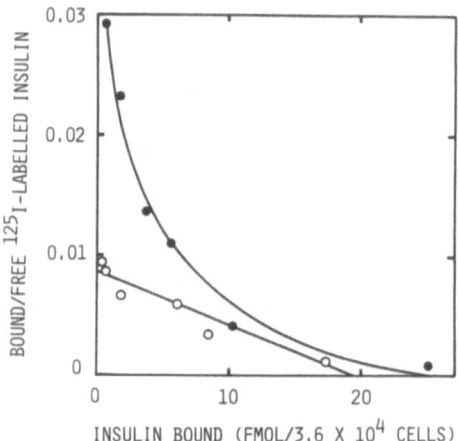

Fig. 1. Scatchard plot of insulin binding to cardiac myocytes in the presence of calcium and magnesium (●–●) or EDTA (○–○). Cells (9 × 10⁴/ml) were incubated in Hepes buffer with calcium (2.5 mmol/l) and magnesium (1.2 mmol/l) or with EDTA (5 mmol/l) for 15 min at 37 °C. Steady-state insulin binding (60 min) was then determined using concentrations ranging from 4.7×10^{-11} mol/l to 2.5×10^{-8} mol/l. All data have been corrected for nonspecific binding and have been taken from four separate experiments. With permission from (13).

Scatchard analysis of equilibrium binding data resulted in a curvilinear plot (Fig. 1) suggesting the presence of more than one class of receptor sites of different fixed affinities (15) or site-site interactions of the negatively cooperative type within a homogeneous class of receptors (18). The K_d of the high-affinity segment was calculated to be 4.5×10^{-10} mol/l with a receptor number of 125 000 sites/cell.

The location of insulin receptors on the surface of cardiac myocytes was elucidated by performing autoradiographic studies. At 37 °C a random distribution was observed with no tendency towards the formation of patches or clusters, as described by Schlessinger for 3T3 fibroblasts (19).

An important role in the binding reaction may be played by the divalent cations calcium and magnesium. Removal of extracellular and presumably membrane-bound calcium by EDTA-treatment resulted in a 60% inhibition of insulin binding. This effect was at least partially reversible by subsequent addition of divalent cations and found to be due to the total suppression of insulin binding to high-affinity sites (Fig. 1). In contrast, the total number of insulin receptors remained unaltered under these conditions. These data suggest that high-affinity insulin receptors are strictly dependent on the ionic environment, supporting the view (20) that calcium may exert a controlling influence on insulin receptor binding.

We have recently characterized the kinetics of 3-O-methylglucose transport in isolated cardiocytes and its modulation by insulin (11). This cellular response has now been used to correlate insulin binding and insulin action and to elucidate the biological function of low-affinity insulin receptors after suppression of high-affinity sites by EDTA-treatment. Under control conditions half-maximal action of insulin was observed at a concentration of 3×10^{-10} mol/l, corresponding to an occupancy of 15% of total receptor sites. These data suggest that the spare receptor concept (1) applies to the heart muscle as well as to other known targets of insulin. EDTA-treatment resulted in a loss of sensitivity of the cardiac cell with half-maximal action of insulin at 10^{-8} mol/l, corresponding to an occupancy of 90% of the low-affinity sites. This absence of spare receptors points to a reduced coupling efficiency, which may be due to the suppression of high-affinity sites or due to the removal of membrane-bound calcium or both. However, in the presence of EDTA an unaltered maximal response was observed which argues in favour of a biological function of low-affinity sites, suggesting that they represent "true" functional receptors for this hormone and are not principally different from the high-affinity sites in this tissue.

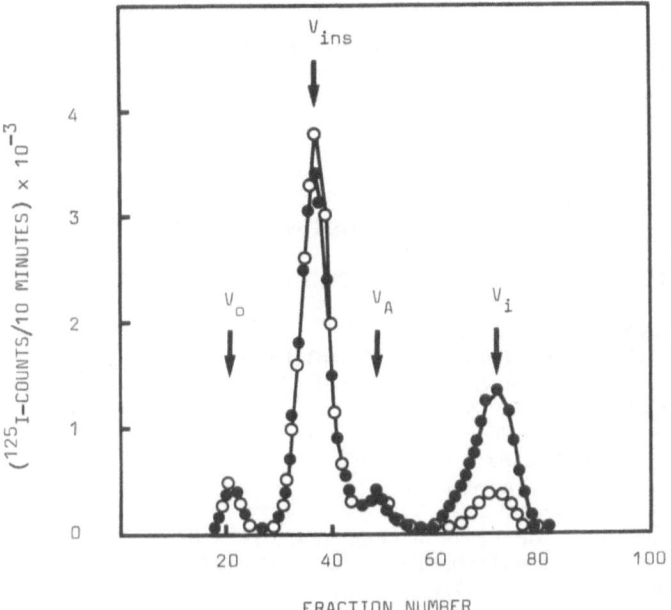

Fig. 2. Gel-filtration profile of radioactivity released from isolated heart cells after chloroquine treatment. Cells were incubated for 30 min in the presence ($O-O$) and absence ($\bullet-\bullet$) of chloroquine (0.1 mmol/l). ^{125}I-labelled insulin was then allowed to associate for 60 min. Separation of free hormone, dissociation of bound radioactivity and gel filtration were performed as described previously (9). With permission from (9).

Receptor-mediated degradation of insulin

Increasing evidence is now available that the degradation of insulin by target cells is mediated by both receptor-dependent and receptor-independent pathways (21). Receptor-independent degradation of insulin by cardiac myocytes has been recently characterized by our laboratory (9). In order to rule out a possible involvement of receptor binding in insulin degradation, we have studied the nature of radioactivity released by cardiac cells which had been allowed to equilibrate with labelled insulin. As can be seen from the gel profile in Fig. 2, more than 40% of bound insulin dissociates in a degraded form, suggesting that insulin bound to the receptor is an efficient substrate for insulin degradation by the isolated heart cell. Receptor-mediated degradation may involve internalization and lysosomal processing (22) or, alternatively, the transfer of insulin molecules to a degrading site in the plasma membrane (21). In order to evaluate these different mechchanisms, we have studied receptor-mediated degradation of insulin after treatment of cells with chloroquine, which is known to be a lysosomotropic inhibitor of proteolysis. As can be seen from Fig. 2, the fraction eluted at the internal column volume, V_i, significantly decreased from 32.2% to 11.0% of total radioactivity in control and chloroquine treated cells, respectively. These observations strongly suggest the involvement of lysosmes in insulin degradation by cardiac myocytes. The physiological implications of this process are still unclear; it may be speculated, however, that after internalization the hormone is degraded to release an active fragment mediating some intracellular response (23).

Acknowledgement

This work was supported by the Ministerium für Wissenschaft und Forschung des Landes Nordrhein-Westfalen, the Bundesministerium für Jugend, Familie und Gesundheit and the Deutsche Forschungsgemeinschaft (SFB 113).

References

1. Kahn CR (1976): J Cell Biol 70: 261–286
2. Freychet P, Roth J, Neville DM, Jr (1971): Proc Natl Acad Sci USA 68: 1833–1837
3. Cuatrecasas P (1971): Proc Natl Acad Sci USA 68: 1264–1268
4. Gavin JR III, Gorden P, Roth J, Archer JA, Buell DN (1973): J Biol Chem 248: 2202–2207
5. Kones RJ, Phillips JH (1975): Cardiology 60: 280–303
6. Eckel J, Reinauer H (1980): Biochim Biophys Acta 629: 510–521
7. Eckel J, Reinauer H (1980): Biochem Biophys Res Commun 92: 1403–1408
8. Eckel J, Offermann A, Reinauer H (1982): Basic Res Cardiol 77: 323–332
9. Eckel J, Reinauer H (1982): Biochem J 206: 655–662
10. Eckel J, Reinauer H (1983): Hoppe-Seyler's Z Physiol Chem 364: 845–850
11. Eckel J, Pandalis G, Reinauer H (1983): Biochem J 212: 385–392
12. Eckel J, Reinauer H (1983): Biochim Biophys Acta 736: 119–124
13. Eckel J, Reinauer H (1984): Diabetes, in press
14. Bar RS, Gorden P, Roth J, Kahn CR, De Meyts P (1976): J Clin Invest 58: 1123–1149
15. Kahn CR, Freychet P, Roth J, Neville DM, Jr (1974): J Biol Chem 249: 2249–2257
16. Duckworth WC (1978): Endocrinology 102: 1766–1774
17. Kahn CR, Baird K (1978): J Biol Chem 253: 4900–4906
18. De Meyts P, Roth P, Neville DM, Jr, Gavin JR III, Lesniak MA (1973): Biochem Biophys Res Commun 55: 154–161
19. Schlessinger J, Shechter Y, Willingham MC, Pastan I (1978): Proc Natl Acad Sci USA 75: 2659–2663
20. Williams PF, Turtle JR (1981): Biochim Biophys Acta 676: 113–117
21. Gliemann J, Sonne O (1978): J Biol Chem 253: 7857–7863
22. Marshall S, Olefsky JM (1979): J Biol Chem 254: 10153–10160
23. Goldfine ID (1977): Diabetes 26: 148–155

Author's address:

Dr. Jürgen Eckel, Diabetes Research Institute, Auf'm Hennekamp 65, D-4000 Düsseldorf 1 (F.R.G.)

Kinetics of cellular uptake of tracers used in myocardial scintigraphy*)

B. Rauch and W. Kübler

Abt. Innere Medizin III (Kardiologie), Universität Heidelberg, Heidelberg (F.R.G.)

Summary

The purpose of this study was to obtain more information about the role of the plasma membrane in cellular uptake of tracers used in myocardial scintigraphy (radioactive thallium chloride and palmitic acid). This knowledge is of importance for an adequate interpretation of the results obtained under clinical conditions.

Key words: thallium, palmitic acid, albumin, myocardial single cells, myocardial scintigraphy

Introduction

In the present study 204-thallium-chloride and 1-14C-palmitic acid have been chosen for investigation. 201-thallium serves as a marker of myocardial blood flow and tissue viability (8). Nevertheless it is not yet fully known to what extent cellular thallium extraction reflects, in addition to myocardial blood flow, function and integrity of the plasma membrane.

1-14C-palmitic acid is used as a marker of cardiac metabolism. Since the oxidation of free fatty acids is reduced in ischemic myocardium, it is possible to detect jeopardized myocardial regions early by diminished fatty acid uptake (6). Apart from cellular metabolism, however, the role of the plasma membrane in fatty acid uptake is still controversially discussed (3, 4, 7, 10).

Methods

To eliminate the influence of regional blood flow, blood vessel wall and interstitial space, myocardial single cells were prepared as follows: hearts of heparinized male Wistar rats were perfused for 10 minutes by a nominally calcium-free buffer solution according to Eckel et al. (2). A second perfusion, lasting 25 minutes, was done using the same solution which also contained 150 U/l crude collagenase and 1000 U/l hyaluronidase. The softened hearts were then sliced in small pieces and reincubated in the same medium for 10 minutes. The resulting cell suspension was washed several times in a solution containing NaCl 130 mM, KCl 4.0 mM, KH_2PO_4 1.2 mM, HEPES 25 mM, glucose 10 mM, and bovine albumin 20 g/l, pH 7.45. After this procedure calcium-sensitive, rod shaped cells were obtained. To overcome the calcium paradox, these calcium-sensitive cells were stored for 15–20 minutes in a medium developed by Isenberg et al. (5), containing high concentrations of substrate and potassium. Afterwards the cells were incubated in the following reaction medium: NaCl 130 mM, KCl 4.75 mM, KH_2PO_4 1.2 mM, $MgCl_2$ 1.2 mM, HEPES 25 mM, glucose 10 mM, essentially fatty acid-free bovine albumin 5 g/l and $CaCl_2$ in stepwise increasing total concentrations of 0.1 mM, 0.5 mM, 2.0 mM and 3.5 mM. With this procedure a smaller part of the cells died. The remaining rod shaped cells (60–80%) tolerated extracellular calcium and showed rhythmic contractions by the external application of current impulses.

---

*) Supported by Deutsche Forschungsgemeinschaft, SFB 90, F.R.G.

To measure the extraction of 204-thallium-chloride by the heart muscle cells a flow dialysis system was used, as described by Colowick et al. (1). Falsification of the results by contamination of the preparations by destroyed cells could be excluded, since cells destroyed by calcium overload were not able to extract 204-thallium-chloride.

Total uptake of 1-14C-palmitic acid was measured by rapid centrifugation of aliquots of the reaction suspension through a layer of silicon oil. Destroyed cells could not pass the silicon oil layer. Extracellular and total space in the pellet was routinely measured by the determination of the 14C-insulin- and 3H-H$_2$O-space.

Unless otherwise indicated, all experiments were done at 37 °C and pH 7.4.

Results and discussion

a) Uptake of 204-thallium-chloride

Fig. 1 shows the time course of thallium net-uptake in resting, calcium-tolerant single heart cells (o; left side). In the presence of 0.5 mM ouabain, which is enough for an almost complete inhibition of the sodium-potassium activated ATPase in the rat heart muscle (9), there was only a small decrease of total uptake within the first 7 minutes (x; left side). A marked reduction of thallium uptake however could be induced by the addition of 5 mM tetraethylammonia (TEA) in the reaction medium (●). This indicates that a passive influx by membrane channels, which are sensitive to TEA, plays a major role in the initial thallium uptake by calcium-tolerant, resting heart muscle single cells.

The situation changed when cells were used which were sensitive to extracellular calcium, due to leakage of the plasma membrane caused by the preparation procedure (Fig. 1; right side). Although these cells were also able to accumulate thallium (intra –extracellular concentration gradient within 7 minutes: 2.3), the ouabain sensitive part of thallium extraction however was markedly increased (x; right side). This could be explained by an increased activity of the sodium-potassium pump compensating the expected increase of thallium efflux due to membrane leakage.

This fact is of importance in clinical scintigraphy, for example, when infarct size reduction has to be evaluated by the application of 201-thallium before and after acute thrombolysis by strep-

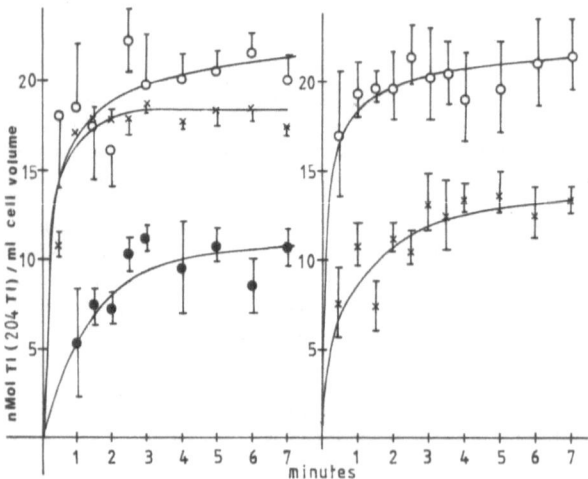

Fig. 1. Time course of 204-thallium uptake in adult rat heart muscle cells without intervention (o), in the presence of ouabain (x) and in the presence of tetraethylammonia (●). Calcium-tolerant cells (left side) were compared with calcium sensitive cells (right side).

tokinase. Early pictures after reperfusion may give an over-estimation of the infarct size reduction due to thrombolysis, since part of the cells which accumulate thallium despite a leaky plasma membrane may additionally die in the following time.

b) Uptake of 1-14 C-palmitic acid

Several studies have been done concerning the uptake of long chain fatty acids in isolated cells of different tissues (3, 7, 10). By most investigators a saturable component in palmitic acid uptake has been described (3, 7, 10). However it is still not clear whether the saturation kinetics account for fatty acid metabolization or membrane transport.

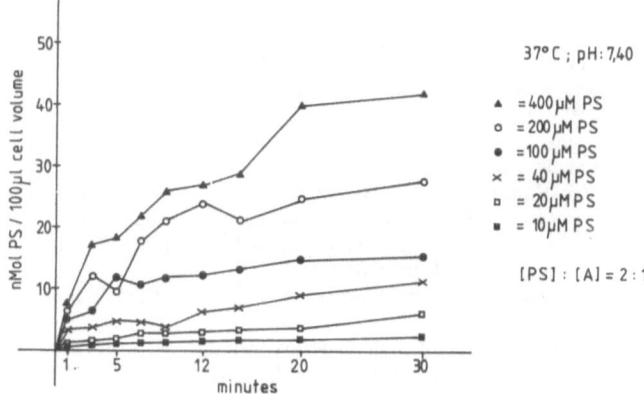

Fig. 2. Time course of 1-14C-palmitic acid uptake in isolated, calcium-tolerant myocytes of adult rat hearts at different total concentrations of palmitic acid as indicated in the figure. The molar ratio between palmitic acid (PS) and albumin (A) was kept constant at 2:1.

Fig. 2 shows the time course of palmitic acid uptake in calcium-tolerant heart muscle single cells at different total concentrations of palmitic acid. The molar ratio of 2 between palmitic acid and albumin was kept constant. As shown by corresponding semilogarithmic plots, the time course of palmitic acid uptake in the first 20 minutes was monophasic. The dependence of the apparent initial rates on total palmitic acid concentration showed saturation kinetics ($V_{max} = 72$ μmol/l · min.; $K_m = 133$ μM).

The saturation kinetics of the apparent initial rates at a constant molar ratio of palmitic acid and albumin could account for intracellular binding sites of enzymes which are involved in fatty acid metabolization. This however seems not to be the case since no saturation kinetics were obtained when the total concentration of palmitic acid was raised at a constant albumin concentration. When albumin was kept constant at 50 μM and the total concentration of palmitic acid was raised from 50 to 250 μM, the relationship between the initial rate and palmitic acid concentration was apparently linear.

From the oxidation rate of palmitic acid in the perfused rat heart and its dependence on substrate concentration, Hütter et al. assumed the existence of an albumin receptor, mediating the cellular uptake of albumin-bound palmitic acid (4). Furthermore, from albumin binding kinetics, the existence of albumin receptors in the liver cell membrane has been assumed (10). The presented data favour the hypothesis of the existence of an albumin receptor in sarcolemma, mediating cellular uptake of albumin-bound fatty acids. However until now, no specific high affinity binding sites for albumin have been demonstrated in heart muscle cells.

Concerning the use of palmitic acid in myocardial scintigraphy, the knowledge of a carrier system as a possible rate limiting factor in the uptake process is of great importance, since in this case palmitic acid could not be further regarded as a metabolic tracer.

References

1. Colowick SP, Womack FC (1969): Binding of diffusible molecules by macromolecules: rapid measurement by rate of dialysis. J Biol Chem 244: 774–777
2. Eckel J, Reinauer H (1980): Characteristics of insulin receptors in the heart muscle. Binding of insulin to isolated muscle cells from adult rat heart. Biochim Biophys Acta 629: 510–521
3. De Grella FR, Light RJ (1980): Uptake and metabolism of fatty acids by dispersed adult rat heart myocytes. J Biol Chem 255, 20: 9731–9738
4. Hütter JF, Piper HM, Spieckermann PG (1984): Kinetic analysis of myocardial fatty acid oxidation suggesting an albumin receptor mediated uptake process. J Mol Cell Cardiol 16, 219–226
5. Isenberg G, Klöckner U (1980): Glycocalyx is not required for slow inward calcium current in isolated rat heart myocytes. Nature 284, No 5754: 358–360
6. Klein MS, Sobel BE (1979): Fatty acid uptake and "metabolic imaging" of the heart. In: Nuclear Cardiology; Willerson JT (ed). FA Davis Company, Philadelphia, pp 165–176
7. Paris S, Samuel D, Jacques Y, Gache Ch, Franchi A, Ailhaud G (1978): The role of albumin in the uptake of fatty acids by cultured cardiac cells from chick embryo. Eur J Biochem 83: 235–243
8. Pitt B, Strauss W (1979): Clinical application of myocardial imaging with thallium-201. In: Nuclear Cardiology; Willerson JT (ed). FA Davis Company, Philadelphia, pp 125–137
9. Repke K, Est M, Portius HJ (1965): Über die Ursache der Speziesunterschiede in der Digitalisempfindlichkeit. Biochem Pharmacol 14: 1785–1802
10. Weisinger R, Gollan J, Ockner R (1981): Receptor for albumin on the liver cell surface may mediate uptake of fatty acids and other albumin-bound substances. Science 211: 1048–1051

Author's address:

Dr. B. Rauch, Abt. Innere Medizin III (Kardiologie), Bergheimerstraße 58, 6900 Heidelberg (F.R.G.)

Adhesion of cardiac myocytes to extracellular matrix components

E. Lundgren, L. Terracio, T. K. Borg

Departments of Anatomy and Pathology, School of Medicine, University of South Carolina, Columbia (U.S.A.)

Summary

The interaction of ECM components with adult cardiac myocytes is not well understood, but is of physiological importance. Most physiological studies are conducted on myocytes in suspension yet *in vivo* the cells are attached to each other and to the ECM. In this paper, we further define the interaction of isolated adult myocytes with the ECM substrates. Of interest is not only the short-term attachment of cells to ECM substrates but also the ability of ECM components to support the long-term maintenance of cardiac myocytes in cultures.

Key words: extracellular matrix, collagen IV, laminin, cell attachment, cell culture

Introduction

The spatial organization of myocytes in the heart is highlighted by cell-cell adhesion as well as cellular adhesion to extracellular matrix components (ECM). Intercellular connections via the intercalated disc and lateral gap junctions have been well described whereas the macromolecular interaction between myocytes and the ECM has only begun to be examined (3, 4, 15, 18, 19).

Interaction with the ECM in vivo

Cardiac muscle is similar to most other organs in the basic organization and types of ECM components present. Components of the ECM are: (1) collagens, including the interstitial collagen types I and III and the basement membrane collagen type IV; (2) glycoproteins such as fibronectin and laminin; (3) proteoglycans like heparin and dermatan sulfate and hyaluronic acid; and (4) new components of yet unknown structure and function.

The connective tissue of the heart is arranged in a 3 dimensional network that is composed of a least 4 components: (1) a weave network that surrounds groups of myocytes; (2) bundles of collagen, 120–150 nM in diameter, that interconnect myocytes; (3) a similar system of collagen that connects myocytes and capillaries; and (4) a system of microthreads that interacts with the collagen bundles (5, 18).

The ECM is intimately associated with the formation of the connective tissue network (2, 3). In the left ventricular free wall of neonates, the connective tissue forms in response to the rapid growth (physiological hypertrophy) following birth. Collagen fibrils, presumably secreted by fibroblasts, form in the ECM and become associated with specific sites on the sarcolemma (3). The arrangement of these ECM components in the heart is similar qualitatively but not quantitatively in most species (5). By 20 days, the connective tissue network in the rat and hamster is morphologically indistinguishable from the adult. Turnover of the ECM components, although not well understood, is probably low unless stimulated as seen in certain types of myocardial hypertrophy where it results in an increase in ECM components.

Interaction of ECM with myocytes in vitro

Cellular recognition and attachment of cells to components of the ECM has been examined in a wide variety of cell types (1, 7, 8, 19) including myoblasts (6, 12) and cardiac muscle (4). Morphologically, collagen attaches to specific sites on the sarcolemma both *in vivo* and *in vitro* (4). *In vitro,* collagen attaches at a specific region on the sarcolemma and actin filaments attach to the same region on the internal face of the membrane (4).

The recognition of ECM components by neonatal and adult cardiac myocytes indicates that there are possible developmental differences in the ability of the myocytes to recognize ECM substrates (4). Neonatal cells readily adhered to fibronectin, collagen types I–V, and laminin whereas adult cells attached only to laminin, type IV collagen, and weakly to fibronectin. The lack of recognition of the interstitial collagen by the adult myocytes indicates that a possible maturation-dependent modulation for the recognition of collagen occurs in the heart. Using an immunological approach, antisera were made against neonatal membranes (anti-N) and against adult membranes (anti-A). Anti-N inhibited the attachment of neonatal myocytes to collagen but not to fibronectin whereas anti-A had no affect. Furthermore, antibodies from hepatocyte membranes that blocked the attachment of hepatocytes to collagen also inhibited neonatal myocytes from attaching to collagen. The anti-N but not the anti-A inhibited the attachment of hepatocytes to collagen. These results indicated that cells of different origins may have similar systems for the recognition of ECM components.

The recognition of the ECM by cells induces a variety of cellular responses including cell spreading (8), contraction (4, 15), lumen formation (11), and control of cell growth (13). Thus the term "receptor" for ECM interaction is justified. These receptors are probably low affinity receptors that cluster during attachment to form a strong connection between the cell and the ECM component (20).

Materials and methods

The methods for isolation of the myocytes has been previously described (15). Briefly, the hearts from 200 g Sprague-Dawley rats were rapidly removed and placed in cold saline. The heart was connected via the aorta to the cannula of a perfusion apparatus and perfused with calcium-free Krebs Ringers Buffer (KRB) containing 0.1% bovine serum albumin (BSA) (KRBA 1) followed by perfusion with 100 µ/ml of collagenase in KRBA 1 for 12 minutes. The heart was then minced, placed in KRBA 2 (KRB containing 2% BSA, 20 µg/ml DNase, 100 U/ml collagenase) and digested on a waterbath-shaker at 37 °C. After 5 minutes released cells were decanted and fresh collagenase was added to the remaining tissue. The digestion was continued for 2 periods of 10 minutes each. Cells released during the last 2 periods were collected by centrifugation and suspended in F-12K medium containing 1.1 mM CaCl₂.

Cell attachment

Substrates of F-12K rat neutral-salt soluble collagen and bovine fibronectin were prepared by previously described procedures (4). Type IV collagen and laminin was purchased from Bethesda Research Laboratory, Bethesda, Maryland, USA. Known concentrations of substrates were added to medium F-12K and plated on Corning multiwell dishes. Substrates were allowed to incubate for a minimum of 30 minutes at 37 °C, rinsed in F-12K, and plated with cells in a known volume of F-12K. Cells were allowed to interact with the substrate for various time periods. The media and unattached cells were aspirated and the dishes washed with F-12K. Adherent cells were fixed with 4% glutaraldehyde in 0.1 M phosphate buffer, pH 7.4 and counted by light microscopy.

Culture procedure

Isolated cells were plated on a known concentration of substrate coated on a glass coverslip, Leighton tube (Costar) or 35 mm Falcon 3002 tissue culture Petri dish. Following adhesion, the media was replaced with F-12K containing 12% fetal bovine serum and placed in an incubator at 37 °C. The cells were fed every 48 hours with F-12K-12% FBS plus 10 µg/ml cytosine arabinocide.

Results and discussion

High yields of calcium-tolerant myocytes were characteristic of the preparations. The same lot of pre-tested collagenase was used for all the initial adhesion studies. The adult myocytes readily adhered to ECM components (Fig. 1). Cells adhered best to basement membrane components of laminin and type IV collagen, followed by fetal bovine serum. Cells were weakly adherent to fibronectin but not to interstitial collagens. Few cells adhered to uncoated plastic but no cells attached to BSA coated plastic. Concentration-dependent adhesion was assayed for all substrates (Fig. 1). A slight plateau occurred for most substrates above 10 μg; however, increased attachment was observed above 40 μg of type IV collagen. Attachment to serum coated plates plateaued above 2.5%, no increased adhesion to fibronectin was seen above 10 μg.

1

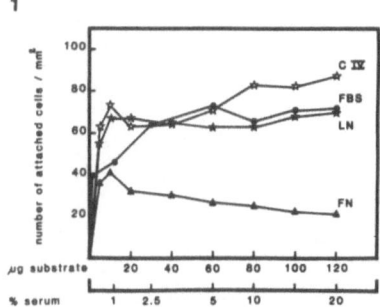

Fig. 1. Attachment of myocytes after 1 hr to various concentrations of collagen type IV (☆), laminin (★), fibronectin (▲), and fetal bovine serum (●).

2

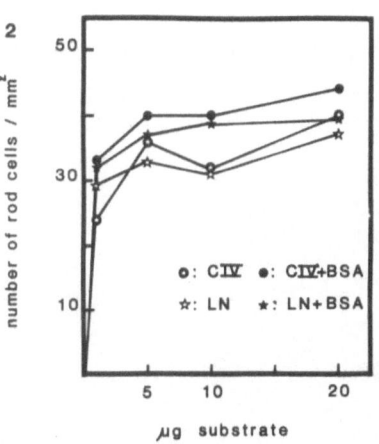

3

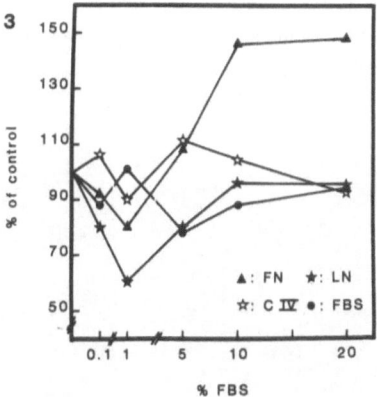

Fig. 2. Effect of BSA on attachment of myocytes after 1 hr to various concentrations of collagen type IV (☆), laminin (☆), collagen type IV + BSA (●), and laminin + BSA (★).

Fig. 3. Effects of various concentrations of fetal bovine serum on attachment of myocytes to collagen type IV (☆), laminin (★), fibronectin (▲), and fetal bovine serum (●). Cells were initially plated with serum and allowed to adhere for 2 hrs. Results with serum added after an initial attachment period of 1 hr are essentially the same. Controls represent attachment to substrates without serum.

The effect of other proteins has been showed to influence the attachment of cells to ECM substrates (9, 10). When dishes were coated with substrates, washed with F-12 K and recoated with BSA no difference in cell adhesion was observed (Fig. 2). The serum concentration has also been shown to influence cell adhesion either before or after cell adhesion (9, 10). Myocytes were exposed to substrates with serum added at the time of plating or 1 hour after cell attachment. Little effect of serum on cell adhesion was apparent except for an increased attachment to fibronectin (Fig. 3). A slight decrease in cell adhesion was seen on laminin at low serum. These data were reproduceable but difficult to explain.

The implications for cell culture of adhered cells are obvious (Figs. 4, 5). A greater than 10 fold increase in the number of adhered and subsequently cultured cells were observed on type IV collagen und laminin (Fig. 4). Although cells readily attached to serum (Fig. 1) their survival in cul-

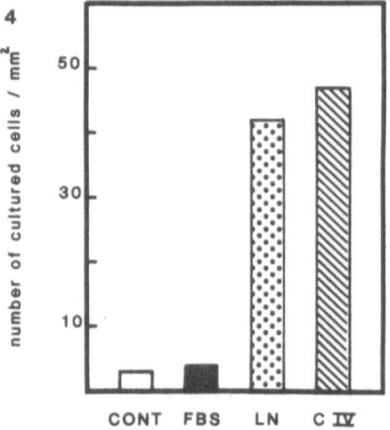

Fig. 4. Number of myocytes after 2 weeks in culture on collagen type IV, laminin, fetal bovine serum, and plastic control.

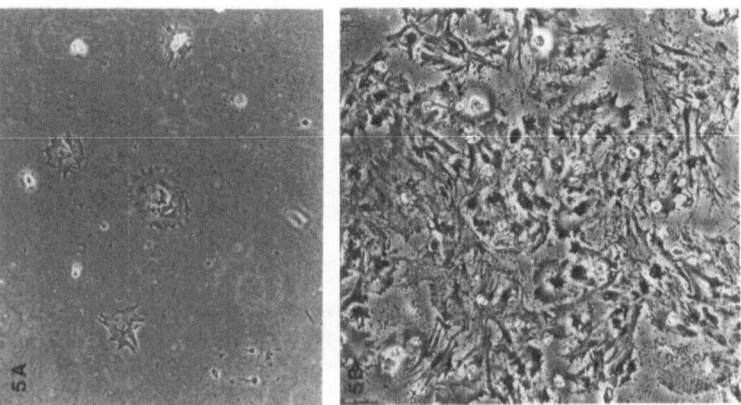

Fig. 5 (a) A phase contrast photomicrograph of myocytes *in vitro* for 2 weeks initially allowed to attach to fetal bovine serum coated plate. (b) Myocytes *in vitro* for 2 weeks initially allowed to attach to collagen type IV.

ture was poor compared to those cells attached to ECM substrates of laminin and type IV collagen. The recognition of these basement membrane components by cardiac myocytes was similar to that observed for other cell types (7, 19). These components are well developed in cardiac muscle and are thought to interact together *in vivo* (7, 14, 16). Separate receptors for laminin have been described in other cells but not in cardiac muscle. Laminin is thought to interact in binding type IV collagen in the basement membrane; however, type IV collagen may have a cell surface receptor independent of laminin. The presence of receptors to the ECM components and their interactions needs further elaboration before more functional studies can be undertaken.

Acknowledgments

This work was supported by grants from the NIH HL-24935 and the American Heart Association #82 912 and 80–714. Intellectual stimulation by K. Rubin and M. Python is gratefully acknowledged. E. Lundgren was supported in part from the Swedish Medical Research Council project 4X–4965 and Kungliga Vetenskapssamhallet, Uppsala, Sweden.

References

1. Aplin JD, Hughes RC (1982): Complex carbohydrates of the extracellular matrix structures, interactions and biological roles. Biochem Biophys Acta 694: 375–481
2. Borg TK, Gay RE, Johnson LD (1982): Changes in the distribution of fibronectin and collagen during development of the neonatal rat heart. Collagen Rel Res 2: 211–218
3. Borg TK, Johnson LD, Lill PH (1983): Specific attachment of collagen to cardiac myocytes *in vivo* and *in vitro*. Develop Biol 97: 417–423
4. Borg TK, Rubin K, Lundgren E, Borg K, Öbrink B (1984): Recognition of extracellular matrix components by neonatal and adult cardiac myocytes. Develop Biol: in press
5. Caulfield JB, Borg TK (1979): The collagen network of the heart. Lab Invest 40: 364–372
6. Chiquet M, Puri EC, Turner DC (1979): Fibronectin mediates attachment of chicken myoblasts to a gelatin-coated substratum. J Biol. Chem 254:5475–5482
7. Couchman JR, Hook M, Rees DA, Timpl R (1983): Adhesion, growth and matrix production by fibroblasts on laminin substrates. J. Cell Biol 96: 177–183
8. Grinnell F (1978): Cellular adhesiveness and extracellular substrata. Int Rev Cytol 53: 67–129
9. Grinnell F, Feld MK (1982): Fibronectin adsorption on hydrophilic and hydrophoboic surfaces detected by antibody binding and analyzed during cell adhesion in serum-containing medium. J Biol Chem 257: 4888–4893
10. Grinnell F, Phan TV (1983): Deposition of fibronectin on material surfaces exposed to plasma: quantitative and biological studies. J Cell Physiol 116: 289–296
11. Hall HG, Farson DA, Bissell MJ (1982): Lumen formation by epithelial cell lines in response to collagen overlay: A morphogenetic model in culture. Proc Natl Acad Sci USA 79: 4672–4676
12. Hauschka SD, White NK (1972): Studies on myogenesis *in vitro*. In: Research in muscle development and muscle spindle: Baker BQ, Prqybyslski RJ, van der Meulen JP, Victor M, (eds) Excerpta Medica, Amsterdam
13. Kleinmann HK, Klebe RJ, Martin GR (1981): Role of collagenous matrices in the adhesion and growth of cells. J Cell Biol 88: 473–485
14. Lesot H, Kuhl U, von der Mark K (1983): Isolation of a laminin binding protein from muscle cell membranes. EMBO J 2: 861–865
15. Lundgren E, Borg TK, Mordh S (1984): Isolation, characterization and adhesion of calcium tolerant myocytes from the adult rat heart. J Mol Cell Card 16: in press
16. Malinoff HL, Wicha MS (1983): Isolation of a cell surface receptor protein for laminin from murine fibrosarcoma cells. J Cell Biol 96: 1475–1479
17. Rao NC, Barsky SH, Terranova VP, Liotta LA (1983): Isolation of a tumor cell laminin receptor. Biochem Biophys Res Commun 111: 804–808
18. Robinson TF, Cohen-Gould L, Factor SM (1983): Skeletal framework of mammalian heart muscle. Lab Invest 4: 482–498

19. Rubin K, Borg TK, Holmdal R, Klareskog L, Öbrink B (1984): Interaction of mammalian cells with collagen. CIBA Found 104: in press
20. Rubin K, Hook M, Öbrink B, Timpl R (1981): Substrate adhesion of rat hepatocytes: Mechanism of attachment to collagen substrates. Cell 24: 463–470
21. Rubin K, Oldberg A, Höök M, Öbrink B (1978): Adhesion of rat hepatocytes to collagen. Exp Cell Res 117: 165–177
22. Sanes JR, Cheney JM (1982): Laminin, fibronectin, and collagen in synaptic and extrasynaptic portions of muscle fiber basement membrane. J Cell Biol 93: 442–451

Author's address:

E. Lundgren, School of Medicine, University of South Carolina, Columbia, South Carolina 29208 (USA)

Development of new intercellular contacts between adult cardiac myocytes in culture

P. Schwartz, H. M. Piper, R. Spahr, J. F. Hütter and P. G. Spieckermann

Zentrum Anatomie und Zentrum Physiologie der Universität Göttingen, Göttingen (F.R.G.)

Summary

On serum precoated tissue culture dishes, isolated ventricular myocytes attach firmly during 4 hours of incubation. Since in this monoculture cells do not divide and show only minimal signs of cytoplasmic spreading, individual cells mostly lie isolated from others. However, when occasionally two cells attach in close vicinity, new cell-cell contact structures are formed already during the first hours in culture. The *de-novo* formation of these communications is demonstrated by the finding that they often connect cells in an end-to-side manner that does not occur *in vivo*.

Key words: gap junctions, cell-cell contact, intercalated disc, cell culture, adult cardiac myocytes

Introduction

During the process of isolation, adult ventricular myocytes change their gross morphology only little. In principle, the separation of the intercalated discs should be the most critical step. Since a method has only recently been described allowing to culture myocytes for a week-span without major morphological changes (4), little is known about the development of the intercalated disc area in isolated cells and about the ability of the cells to form new intercellular contacts *in vitro*.

It has been previously shown that cultured myocytes become smooth at their intercalated discs during the first 2 days *in vitro* (4). After that time, the sarcolemma at this site becomes indistinguishable from that of the lateral cell surface by conventional electron microscopy. It is demonstrated in this study that only spatial contact to an adjacent cell is needed for the formation of new intercellular contacts *in vitro*.

Methods

Ventricular muscle cells were isolated from 12 weeks old female Sprague-Dawley rats and plated on 60 mm dishes in M-199 medium containing 4% fetal calf serum, as previously described (4). Thereafter, serum was omitted. M-199 medium was supplemented with 5 mM creatine and 10^{-9} M insulin. Medium was changed every 24 hours.

For transmission electron microscopy, cells were fixed in 2% glutaraldehyde in 0.1 M cacodylate buffer at pH 7.3. After 2 hours fixation at 4 °C, the cells were washed in buffer and postfixed in cacodylate-buffered 2% osmium tetroxide. Then specimens were embedded in Araldite after dehydration. Ultrathin sections were stained with 2% methanolic uranylacetate and lead citrate and examined in a Zeiss EM 10 microscope (Zeiss, Oberkochen, Germany). For scanning electron microscopy cells were fixed in the same way. After dehydration in graded ethanols using amylacetate as intermedium, samples were dried in a critical point dryer, and coated with gold/palladium. Then specimens were examined in a Novascan 30 (Zeiss).

Results and discussion

Freshly isolated cells have sharp polygonal contours with sponge-like membrane foldings at their intercalated disc areas. The fact that these surface areas retain their augmented structure, even in the absence of external forces, indicates the existence of a rigid cytoskeletal support. Obviously most intercellular contacts are cleaved symmetrically, however, frequently vesicular membrane particles are observed which adhere to the plasmalemma with a connection typical for a gap junction (1, 5). Occasionally little vesicles with a double membrane exhibiting the same type of intermembraneous contact are also seen inside the cell body near an intercalated disc area. Such observations have been interpreted as incorporations of gap junction structures torn out of the plasma membrane of an adjacent cell (5). If this is the major route for disconnecting cells at the gap junction contacts, then this tearing-out in general cannot be very harmful to the cells since otherwise not as much as 30 percent of the ventricular cell mass might survive cell separation. Because of the multitude of gap junctions per cell (3), the existence of such cells which do not lose a single gap junction structure to the advantage of an adjacent cell is highly improbable.

After 12 h in culture, the parasarcolemmal densifications characteristic for specific contact structures have disappeared and the depth of the membrane foldings at the former intercalated discs is greatly reduced. After 24 hours neither adherent nor incorporated gap junction structures are seen in cultured cells. After 2 days the cell endings are round and smooth, covered with a continuous ruthenium-red stainable surface coat.

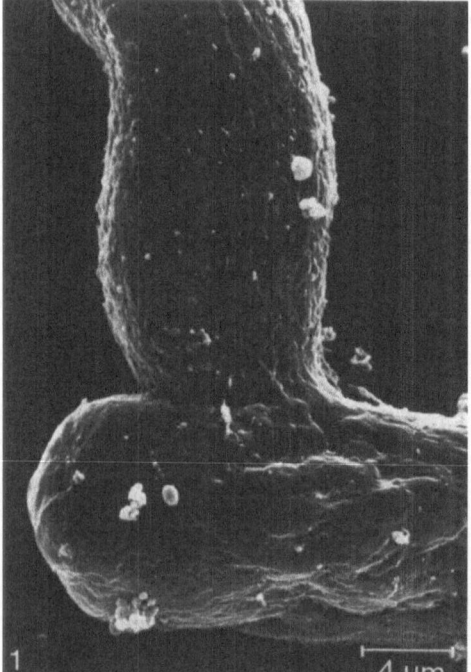
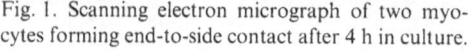
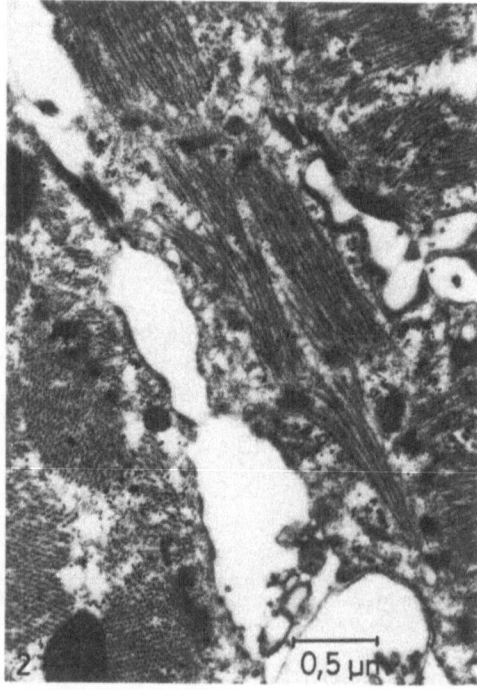

Fig. 1. Scanning electron micrograph of two myocytes forming end-to-side contact after 4 h in culture.

Fig. 2. Transmission electron micrograph of an intercellular contact area on day 3 in culture. The cell on top, right, contains a sarcolemmal enclosure in which bridges of thin material connect opposing surface areas.

Fig. 3. Transmission electron micrograph of an intercellular contact area on day 4 in culture. There are two gap junction-like structures, the one on top is accompanied by two parallely oriented SR tubules.

Early cellular attachment (3 h) of rod-shaped cells is mediated by the development of small cytoplasmic projections at sarcomeric distances anchoring the cell bottom to the petri dish surface. Similar protrusions are not visible at the upper part of the cells, indicating a polarized specialisation of the sarcolemma. Large pseudopodia extending from the cell endings are often seen in cells after 3 days (4).

Although intercellular distances and absence of cell spreading in this system prevent development of new cell-cell contacts in culture for most individuals, occasionally cell cluster formation favours their formation. Indeed, already after 4 h, adjoining cells form contacts in the form of small cytoplasmic bridges with membrane thickening at opposing contact areas and with intermembraneous material deposits. In most cases these contacts are end-to-side, which is the most probable mutual orientation of two randomly oriented rods (Fig. 1). The fact that such end-to-side connections do not occur in heart tissue proves the *de-novo* character of these contacts. Later, more extended contact structures are found. An example from day 3 in culture is given in Fig. 2. Again myofibrils of both cells are not in line. Probably due to a rolling-in of cell edges at the cell endings, which is often seen during the rearrangement period (4), a sarcolemmal enclosure is seen (top right) in which opposite sarcolemmal faces of the same cell form thin contacts.

In the further development, cellular contact areas with several specific contact structures extend to larger areas. Fig. 2 shows a detail from such a 'primitive intercalated disc' on day 4. According to the intermembraneous distances both contact structures may represent gap junctions (3). On top the close spatial vicinity of both plasmalemmas is accompanied by two parallely oriented straight SR tubules. These observations suggest that isolated cells again may gain low conductivity contacts. In contrast to the case of cell cultures of postnatal origin in which most cells start spontaneous beating on contact formation (2), beating clusters were not observed during a week in culture.

It is demonstrated by this study that sarcolemmal specialisations mediating cell contact *in vivo* can be regained by isolated cells *in vitro*. At present it is not known if these structures undergo a disassembly-reassembly cycle or are degraded and newly synthesized. It is remarkable that close plasmalemmal contact seems to be sufficient for inducing contact structure formation on any part of the cell surface. If one could further improve seeding density a monolayer network might be obtained, which again exhibits functional and trophic cell-cell interactions.

Acknowledgement

This study was supported by the Deutsche Forschungsgemeinschaft, SFB 89, Kardiologie Göttingen. The assistance of B. Eickhoff, H. Haacke and R. Zöllner is greatly acknowledged.

References

1. Flegler-Baron C, Behrendt H (1982): Effects of Ca^{2+} deficiency, collagenase, and mechanical dispersion on the ultrastructure of cardiac myocytes of adult rats. Eur J Cell Biol 27: 262–269
2. Harary I, Farley B (1960): In vitro organisation of single beating rat heart cells into beating fibres. Science 132: 1839–1840
3. McNutt NS (1970): Ultrastructure of intercellular junctions in adult and developing cardiac muscle. Am J Cardiol 25: 169–183
4. Piper HM, Probst I, Schwartz P, Hütter JF, Spieckermann PG (1982): Culturing of calcium stable adult cardiac myocytes. J Mol Cell Cardiol 14: 397–412
5. Severs NJ, Slade AM, Powell T, Twist VW, Warren RL (1982): Correlation of ultrastructure and function in calcium-tolerant myocytes isolated from the adult rat heart. J Ultrastruct Res 81: 222–239

Author's address:

P. Schwartz, Zentrum Anatomie, Universität Göttingen, Kreuzbergring 36, D-3400 Göttingen (F.R.G.)

Long-term primary cultures of adult human and rat cardiomyocytes

S. L. Jacobson, M. Banfalvi, and T. A. Schwarzfeld

Department of Biology, Carleton University, Ottawa (Canada)

Summary

An improved protocol for isolating and culturing adult mammalian cardiomyocytes is presented. Problems of establishing and maintaining cultured adult human and rat cardiomyocytes, and some of their properties, are discussed.

Key words: adult, human, rat, cardiomyocyte, culture

Introduction

Five laboratories have published data on cultured adult rat cardiomyocytes from ventricular cells (3, 4, 8, 9) and atrial cells (2). Extensive inference about the value of such preparations is premature; however, the future appears bright. Adult myocardium with its tightly-coupled structure is unlikely to yield undamaged cells. But, in an appropriate culture system, cells could repair damage, and cultures could provide a system for production and manipulation of large homogeneous populations of normal cardiomyocytes.

Our laboratory described a method for isolating and culturing adult cardiomyocytes (4). Since then cell yield has been increased by about two orders of magnitude and survival of myocytes in culture has increased to about 75%. We have cultured cells from the adult mouse, hamster, guinea-pig and monkey, and from the rat and the human. We have concentrated on improving and studying cultures of rat cells. However, we also have results with human cardiomyocytes in culture. Some of those data are reported here.

Materials and methods

Standard salines: Two buffered salines (pH 7.2–7.4) based on data from (1) are used in the isolation procedure. These are (in mM): Saline 1: NaCl 90, KCl 30, NaHCO$_3$ 2, glucose 5.5, sucrose 42, HEPES 2, phenol red 20 mg/l. Saline 2: NaCl 120, KCl 5, KH$_2$PO$_4$ 1.2, Na$_2$HPO$_4$ 16, MgCl$_2$ 1.2, glucose 11, phenol red 20 mg/l.

Preparative procedure: Cells are derived from the heart of a male Sprague-Dawley rat, 3 to 4 months old. The rat is decapitated with a guillotine and the apical one-half of the heart is excised,rinsed in saline 1 and minced into 2 to 3 mm pieces with a razor blade or scissors. Tissue is transferred to a specially designed disaggregation vessel (DV) containing 8 ml of 0.25% (W/V) trypsin, (Sigma Chemical Co. type XI), in saline 1. The DV is as described in (4) except that it is larger with an O. D. of 50 mm. The DV is kept at 37 °C during cell isolation and the DV fingers are adjusted for full-length interdigitation. Tissue is treated initially for 7 min with the DV shaft rotated at 8 RPM. Afterwards trypsin is removed and discarded. Eight ml of 0.1% collagenase, (Sigma ChemCo. type I or IA), in saline 1 are then used to treat the tissue at 8 RPM for 15 min. The collagenase containing isolated myocytes is then centrifuged at 70 G for 10 min at 21 °C. The centrifuged collagenase is decanted leaving a loose pellet of cells (harvest 1) which is gently resuspended in 1.5 ml of saline 2 and stored at 37 °C. The DV is refilled with 8 ml of fresh 0.1% collagenase for a second 15 min treatment. Treatment of the tissue continues with 0.1% collagenase for three harvests and subsequently for four more harvests using 0.05% collagenase. Cells from each harvest are resuspended in an aliquot of the same 1.5 ml of saline 2 and stored, pooled, at 37 °C. The resuspension will contain myocytes, non-myocytes and debris.

The resuspension is treated in one of 3 ways to establish the myocytes in culture: (i) brought to a desired number of cells/ml by addition of growth medium, i.e. medium 199 (M199) plus 20% fetal bovine serum, and plated; (ii) density sedimented on 6% bovine serum albumen (BSA), (Sigma Chemical Co. A 9647), in M199 at 5 G for 10 min at 21 °C. A layer of intact cells will be visible near the bottom of the tube and can be recovered by aspiration, resuspended in growth medium, and plated. BSA sedimentation gives intact myocytes devoid of about 90% of non-myocytes and debris; (iii) the resuspension, BSA sedimented or not, can be brought to an appropriate volume with growth medium containing 10 µM cytosine 1-β-D-arabinofuranoside (ARA), (Sigma Chemical Co. C 6645), and plated for 7 days. On day 7 the medium is changed to medium without ARA. This treatment eliminates nearly 100% of non-myocytes after about three weeks in culture.

Results

Rat cardiomyocytes: Data are given on development of these cells in culture in (10) and on their electrophysiological properties in (6). We are examining DNA synthesis in our cultures and a preliminary report of these data will be presented at this meeting (5).

To review briefly, we and others (3, 8) observed that isolated adult cardiomyocytes in culture undergo a stereotypic morphologic change. Elongated newly isolated cells gradually become round. This change can take several days (10). Often the first sign of rounding is bulging at the end(s) of the cells. Cells attach to the substratum after 24 to 72 hours, apparently coincident with the onset of rounding. Attached cells extend processes. Their spreading and flattening follow directly and are complete by 10 to 20 days in culture. The rate of morphologic change is slower in cultures treated with ARA. These cultures can lag non ARA cultures by as much as a week.

The electrophysiological state of our cultured cells is depolarized, with functional Na^+ and Ca^{++} channels (6). We are investigating the cause of depolarization. We are also studying membrane polarization with time in culture. We know that at least some newly isolated cells are normally polarized (-80 to -90 mv) and that by 9 days in culture the polarization is about -25 mv. We lack data from 1 to 9 days, but presumably the cells depolarize within that period. Our data also indicate that polarization rises from 9 to 60 days reaching about -70 mv.

Two groups (3, 8) report cultured ventricular cells that actively synthesize DNA. One of these published details (3). In our cultures, mitosis of myocytes can be detected only if cells are from rats younger than 23 days (10). Furthermore data indicate (5) that the rate of ^{3}H-Thymidine uptake in our cultures is considerably less than that reported in (3).

Human cardiomyocytes: Our laboratory is ten minutes' drive from a hospital where heart surgery is performed. In surgery requiring cardio-pulmonary bypass, a catheter is placed in the right atrium. The catheter passes via an opening made by excising part of the right atrial appendage. The excised tissue is normally discarded. Papillary muscle excised in valve replacement surgery can also occasionally be obtained. We used right atrial appendage as a source of cells because it could be obtained regularly.

At first cells were treated according to the protocol in (4). Subsequently cells were prepared with variations of this protocol, including one similar to that outlined in the Materials and methods section. Several culture media including M199 and minimal essential medium (MEM) and various sera including bovine fetal and calf were tested. Several ways to maintain tissue during the approximately 20 min between excision and beginning cell isolation were also tested. Tissue was maintained in lactated Ringers, M199, and MEM, both oxygenated and aerated, on ice and at ambient temperature. It appears that the quality of serum used in the growth medium and the number of cells obtained in the isolation procedure are the most important factors in establishing human cells in culture. When serum known to be efficacious in rat cell cultures was used and when a large number of cells were isolated in apparently good condition, they established in culture irrespective of the other parameters. In the first 10 to 14 days in culture, the human cells behaved in a way similar to that described for rat ventricular cells (10). After that, cells appeared to lose their myocytic properties (contractility and striated morphology).

Figure 1 shows excerpts from a phase contrast photomicrographic series of a human cell culture. The cells were isolated from the right atrial appendage of a 58 year old male who underwent coronary artery bypass graft surgery. The photo series was obtained with a numerically-controlled microscope and ancillary instrumentation described in (10). Three myocytes are shown. All eventually developed in culture to the spread stage. One cell (arrow) developed most rapidly (A); rounded (B); spread, acquiring striations and contracting (C–G). Eventually it lost its apparent myocytic properties (H).

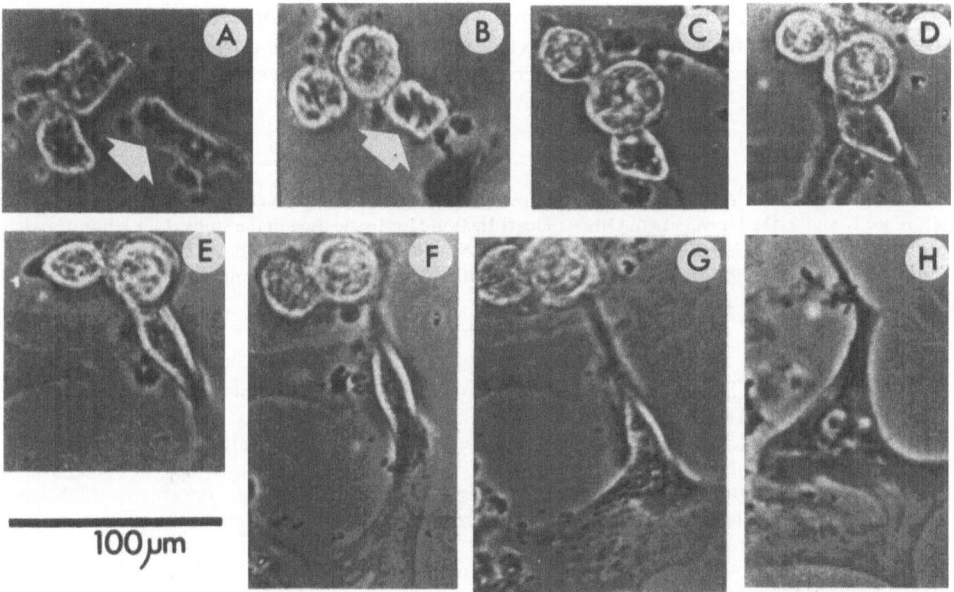

Fig. 1. Excerpts from a time-lapse sequence showing development of isolated human cardiomyocytes in culture. (Frame/days: hours) A/1:3, B/1:20, C/7:14, D/7:23, E/8:11, F/9:9, G/9:12, H/11:0.

Discussion

Other laboratories will want to use cultured adult cardiomyocytes in the near future. Using poorly-defined systems to determine the unknown has obvious pitfalls. Defining properties of these cells and defining the factors which determine them are prerequisites for maximum scientific utility, greater quality, and simplicity of culture preparation.

Isolated adult cells are very fragile compared with neonatal cells. They are intolerant of sedimentation, stirring, titurating or other ungentle treatment. They are also very sensitive to chemical factors which may appear in autoclave steam, laboratory air or water supplies, and materials (enzymes, sera, etc.) that come into close or long-term contact with cells or cultureware. Problems with such factors have cost us many months of difficulty in the laboratory. Materials must therefore be tested and selected for their efficacy.

We have tried many things which did not work well. Some may be of interest to others. Trypsin or collagenase alone, with our mechanical treatment, fails to produce cells. Exposure to trypsin must precede, or accompany, exposure to collagenase for acceptable results. As with human cells, an efficacious serum for rat and other cells is of great importance. The elimination or reduction of serum used is desirable. We observed that it has an effect on spontaneous contractility of cells (6). NuSerum (a product of Collaborative Research, Inc.) works as a serum substitute in our cultures. However, it contains 25% bovine serum and is a secret formulation, making it an undesirable, if not unsuitable, substitute for another unknown – i. e., natural serum. We have used rat serum for rat cells and also used serum prepared from the blood of the rat whose heart gave the cells (unpublished data). This proved to be an inadequate substitute for bovine serum unless mixed with it. Lastly, we tried Percoll and Ficoll in place of BSA as a density sedimentation medium. Both damaged the cells.

There are reports of cultured human embryonic cardiac explants, of isolation of adult human cardiomyocytes, and a few reports of cultured human embryo cardiomyocytes (cf. review in 7). However, we could find no report of success in culturing adult human cardiomyocytes. In the future, such cultures may offer a way not otherwise ethically possible to study details of human cellular cardiology.

Acknowledgements

We are grateful to the Merrell Dow Research Institute, Indianapolis, Indiana for support of the human cell culture study and to the University of Ottawa Heart Institute for tissue used in the study. Our laboratory has been supported by the Carleton Faculty of Graduate Studies and Research and the Ontario Heart Foundation. We thank Mr. Cliff Wilson and the personnel of the Carleton Science Technology Centre for help with the design and fabrication of instrumentation.

References

1. Altschuld R, Gibb L, Ansel A, Hohl C, Kruger FA, Brierley GP (1980): Calcium tolerance of isolated rat heart cells. J Mol Cell Cardiol 12: 1383–1395
2. Cantin M, Ballak M, Beuzeron-Mangina J and Anand-Srivastava MB (1981): DNA Synthesis in cultured adult cardiocytes. Science 214: 569–570
3. Claycomb WC and Bradshaw HD Jr (1983): Acquisition of multiple nuclei and the activity of DNA polymerase α and reinitiation of DNA/replication in terminally differentiated adult cardiac muscle cells in culture. Dev Biol 99: 331–337
4. Jacobson SL (1977): Culture of spontaneously contracting myocardial cells from adult rats. Cell Struct and Funct 2: 1–9
5. Jacobson SL: DNA Synthesis in cultured ventricular cells of the adult rat. Presented at the International E. Riesch Symposium on Isolation, Properties and Applications of Adult Heart Cells. Spangenberg Castle, Federal Republic of Germany, March 22–25, 1984
6. Jacobson SL, Kennedy CB, Mealing GAR (1983): Evidence for functional sodium and calcium ion channels in the membrane of cultured cardiomyocytes of the adult rat. Can. J. Physiol. and Pharmacol. 61: 1312–1316
7. Liebermann M, Adam WJ and Bullock PN (1980): The cultured heart cell: Problems and Prospects. In: Methods in Cell Biology, Vol 21 A, Academic Press, New York, pp 187–203
8. Nag A and Cheng M (1981): Adult mammalian cardiac muscle cells in culture. Tissue and Cell 13: 515–523
9. Piper HM, Probst P, Schwartz P, Hutter FJ, Spieckerman PG (1982): Culturing of calcium stable adult cardiac myocytes. J Mol Cell Cardiol 14: 397–412
10. Schwarzfeld TA and Jocobson SL (1981): Isolation and development in cell culture of myocardial cells of the adult rat. J Mol Cell Cardiol 13: 563–575

Author's address:

S. L. Jacobson, Department of Biology, Carleton University, Ottawa, Ontario K1S 5B6 (Canada)

Morphological dedifferentiation of adult cardiac myocytes in coculture with hepatocytes

R. Spahr, H. M. Piper, P. Schwartz, I. Probst, and P. G. Spieckermann

Zentrum Physiologie, Zentrum Anatomie und Zentrum Biochemie der Universität Göttingen, Göttingen (F.R.G.)

Summary

When adult heart cells are plated on a dish covered with a monolayer of hepatocytes gradual morphological changes are observed. While during the first day the myofibrils are still organized in rod-like shape, later the cells become flat and spread on top of underlying hepatocytes. After two days most cells have a flat, polygonal appearance with spread myofibrillar bundles. At this stage they start spontaneous rhythmic contractions which are characteristic for embryonic myocytes, but not for isolated adult ventricular cells. In this culture myocytes form specific contact structures to adjacent myocytes as well as to hepatocytes. These results demonstrate the phenotypical plasticity of adult heart muscle cells which are believed to be terminally differentiated.

Key words: dedifferentiation, cell spreading, cell-cell contact, hepatocytes, adult cardiac myocytes

Introduction

We have reported previously that isolated ventricular muscle cells from adult rat hearts can be held for a week-span in culture, where they undergo only minor morphological alterations (5). Thus, myofibrils remained in register and signs of spreading were confined to the development of single pseudopodia at the cell poles. With the applied culturing technique, an attachment of the cells could be obtained during the first 3 hours. In this respect, our culture system differed in principle from those described by others (1, 4, 7) in which early attachment was not achieved. In those, nonattached cells rounded off during the first days, myofibrils were disintegrated and mitochondria showed severe signs of dissolution. However, these features were regarded as an intermediate stage for a new state of cellular differentiation. This is because after a few days in culture, cells exhibiting properties of an earlier stage of development were found attached to the culture dish surface. Like heart cells of immature type these cells were spontaneously beating, showed extensive cellular spreading and contained myofibrils in diverging orientations. In the hepatocyte-myocyte coculture, described here, a morphological dedifferentiation of adult myocytes develops gradually during the first days in culture without a round intermediate stage.

Methods

Hepatocytes were isolated from 12 weeks old Sprague-Dawley rats and plated as described elsewhere (6). Ventricular muscle cells from the same rat strain were isolated as previously described (5) and plated on top of the 4 hour old hepatocyte monolayer. After an attachment phase of 4 hours with 4% fetal calf serum, serum was omitted and the culture was maintained in M-199 medium with 10^{-9} M dexamethasone and 10^{-9} M insulin.

For transmission electron microscopy, cells were fixed in 2% glutaraldehyde in 0.1 M cacodylate buffer at pH 7.3. After 2 hours fixation at 4 °C, the cells were washed in buffer and postfixed in cacodylate buffered 2%

osmium tetroxide. Then specimens were embedded in Araldite after dehydration. Ultrathin sections were stained with 2% methanolic uranylacetate and lead citrate and examined in a Zeiss EM 10 microscope (Zeiss, Oberkochen, Germany). For scanning electron microscopy cells were fixed in the same way. After dehydration in graded ethanols using amylacetate as intermedium, samples were dried in a critical point dryer and coated with gold/palladium. Then specimens were examined in a Novascan 30 (Zeiss).

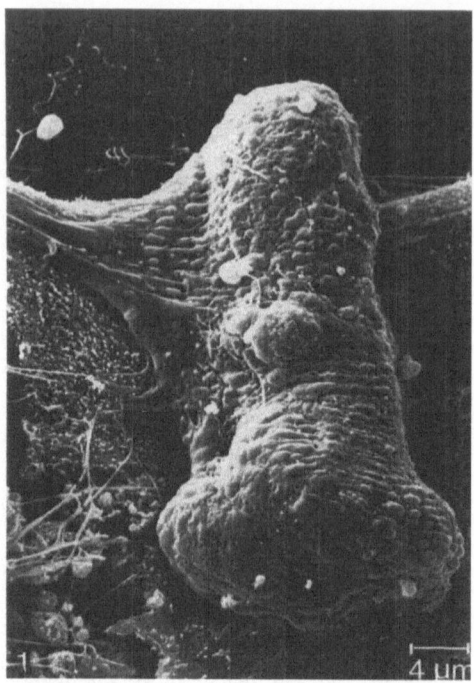

Fig. 1. Scanning electron micrograph of a myocyte on day 3 in coculture with hepatocytes.

Results and discussion

During the first day of culture, myocytes lie in random orientation on the hepatocyte substratum, still exhibiting the typical elongated shape of freshly isolated cells. However, the former intercalated disc area already becomes smooth and round at its edges. In fenestrations of the hepatocyte substratum myocytes often gather in clusters. After 2 days, cell endings of the myocytes are either rounded butt-like or flattened and spread out. In the myocyte shown in Fig. 1, myofibrils are still organized in one voluminous bundle, while thin sarcolemmal sheets are spread out over the underlying hepatocytes. Hepatocyte surfaces can be recognized by their microvilli. The cell ending in front is thickened, as is also often observed during the reorganization process of the intercalated disc areas of myocytes in myocyte monocultures (5). A thin sheet of cytoplasm is expanded from the bottom of the cell pole seen in front. At the lateral cell surfaces the cytoplasm is spread out like a wing. The thin sarcolemmal sheet seems to be stretched out over mitochondria which have marched out of the cellular body at sarcomeric distances.

After 3 days on the hepatocyte monolayer most myocytes are flattened and spread out in a large polygonal shape. Thus, under light microscopy, their presence is often only recognized by their cross-striations. Myocyte clusters often exhibit a network appearance. At this time, many myocytes spontaneously start beating, with frequencies up to 300 beats/min. Beating, however, occurs mostly in bursts with intervening silent periods. In cell clusters beating is often synchronized in a multitude of cells (3). In contrast, in a myocyte monoculture in which these cells undergo only minimal morphological changes, even after one week beating can only be elicited by electrical stimulation (5, 9).

During the development of structural changes myofibrillar parallelity is first lost. On day 2, myofibrils often appear distorted. However different from similar orientations in hypercontracture, these myofibrils are still relaxed. After day 3, many cells exhibit disintegrated myofibrillar material with dissolution of Z-bands. However, even in these cells morphological indications of a well preserved energetic state are present: glycogen is abundant and has even markedly increased during the culture period, and mitochondria contain small matrix granules seen only at high reserves of creatine phosphate (8).

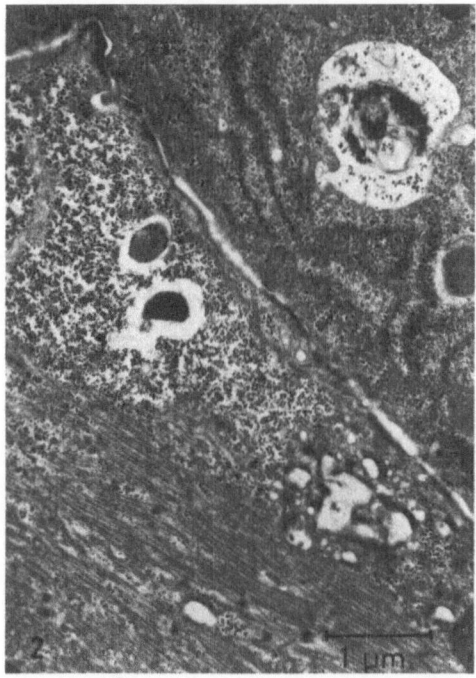

Fig. 2. Transmission electron micrograph of a hepatocyte-myocyte contact area on day 4 of the coculture. The hepatocyte covers the picture on top right. Both cells contain abundant glycogen.

After 1 and 2 days in culture, first signs of contact formation between myocytes and heptocytes are observed: small cytoplasmic projections reach the other cell's surface. After 3 days in culture, broad contact areas are formed (Fig. 2). Specific contact structures can be recognized by their coincident parasarcolemmal densifications on both plasmalemma faces. At the same time, extended contact areas are also present in myocyte clusters. While it is difficult to categorize the nature of the immature contact structures formed between heptocytes and myocytes at this stage, some of those formed between myocytes are apparently gap junctions. This coincidence suggests a causal relation between gap junctions and beating synchronicity (2), even if beating synchronicity may not always depend on the existence of gap junctions (10).

These observations demonstrate that adult heart muscle cells, under appropriate environmental conditions, may pass through a continuum of morphological alterations. During this development they regain features of earlier developmental stages. Thus, spontaneous beating is character-istic for cultured embryonic or postnatal cells (3). In agreement with the report of Claycomb and Bradshaw (1) we did not observe cell divisions of myocytes during one week in culture. It is not yet known which factors in this coculture are responsible for the phenotypical changes of heart muscle cells. However, the influence of hormones or growth factors supplied by the hepatocytes might be a likely cause. The phenotypical plasticity of adult heart cells demonstrates that under appropriate environmental conditions the cellular elements of the myocardium can undergo a variety of structural alterations apart from hyper- and atrophia or degenerative changes which are observed *in vivo*.

Acknowledgements

This study was supported by the Deutsche Forschungsgemeinschaft, SFB 89 – Kardiologie Göttingen. The assistance of B. Eickhoff, H. Haacke and R. Zöllner is greatly acknowledged.

References

1. Claycomb WC, Bradshaw HD (1983): Acquisition of multiple nuclei and the activity of DNA polymer-ase α and reinitiation of DNA replication in terminally differentiated adult cardiac muscle cells in culture. Dev Biol 99: 331–337
2. Griepp EB, Peacock JH, Bernfield MR, Revel JP (1978): Morphological and functional correlates of synchronous beating between embryonic heart cell aggregates and layers. Exp Cell Res 113: 273–282
3. Harary I, Farley B (1960): In vitro organisation of single beating rat heart cells into beating fibres. Science 132: 1839–1840
4. Nag A, Cheng M (1981): Adult mammalian cardiac muscle cells in culture. Tissue and Cell 13: 515–523
5. Piper HM, Probst I, Schwartz P, Hütter JF, Spieckermann PG (1982): Culturing of calcium stable adult cardiac myocytes. J Mol Cell Cardiol 14: 397–412
6. Probst I, Schwartz P, Jungermann K (1982): Induction in primary culture of 'gluconeogenic' and 'glyco-lytic' hepatocytes resembling periportal and perivenous cells. Eur J Biochem 126: 271–278
7. Schwarzfeld TA, Jacobson SL (1981): Isolation and development in cell culture of myocardial cells of the adult rat. J Mol Cell Cardiol 13: 563–575
8. Schwartz P, Piper HM, Spahr R, Spieckermann PG (1984): Ultrastructure of cultured adult myocardial cells during anoxia and reoxygenation. Am J Pathol 115: 349–361
9. Schwartz P, Piper HM, Spahr R, Hütter JF, Spieckermann PG (1984): Development of new intercellular contacts between adult cardiac myocytes in culture. Basic Res Cardiol 79, suppl. 2: 75–78
10. Sperelakis N (1979): Propagation mechanisms in heart. Ann Rev Physiol 41: 441–457

Author's address:

R. Spahr, Zentrum Physiologie, Universität Göttingen, Humboldtallee 23, D-3400 Göttingen (F.R.G.)

Electrophysiological properties of isolated ventricular myocytes

T. Powell

Department of Medical Physics and Institute of Nuclear Medicine, The Middlesex Hospital Medical School, London (U.K.)

Summary

There is no doubt that major advances have been made in cardiac electrophysiology using single heart cells. Accurate analysis of rapid inward sodium current and the second inward current carried by calcium has resulted in a major re-examination of the steps involved in excitation-contraction coupling in the heart. Future work using glass microelectrodes, suction pipettes and patch pipettes will yield a vast amount of information highly relevant to mechanisms involved in the initiation of the heartbeat.

Key words: electrophysiology of isolated ventricular myocytes, electrical characteristics, voltage clamp

Introduction

The availability of individual heart cells has had an immediate and important impact in the field of cardiac electrophysiology. With the whole heart, difficulties stem from the intricate anatomy of the myocardium, coupled with distinct morphology and there are problems in the interpretation of experimental data obtained from the whole organ. The simplified geometry of the single cell, with the lack of restricted intercellular, interstitial and intravascular spaces, together with the possibility of improved spatial and temporal control in voltage clamp experiments, have prompted much activity in recording electrophysiological data from isolated myocytes.

Passive electrical characteristics

An imperative pre-requisite for studies using isolated cardiac cells is that the resting electrical characteristics of the myocytes compare favourably with those recorded from multicellular preparations. Initial investigations of the electrical properties of individual rat ventricular cells were not encouraging, since action potentials recorded from enzymically released cells evolved from membrane potentials of -30 to -50 mV (8, 9). Our own early work highlighted the difficulties inherent in recording stable healthy resting potentials from rat cells which were otherwise quiescent and calcium-tolerant (22), but shortly afterwards we reported resting membrane potentials in the range -70 to -90 mV for superfused cells at 37 °C and described briefly the effects of adrenaline on electrical activity (23). In contrast, another report a year later in 1979 maintained that no resting potentials more negative than -10 mV, nor any action potentials, could be measured from isolated rabbit myocytes (26). We then published comprehensive results on the electrical properties of rat cells in 1980 (24) and concluded that the isolated myocytes could maintain resting potentials more negative than -70 mV and retained the ionic mechanisms to support action potentials. Although this paper (24) has proved a benchmark for single cell electrophysiology, some of the results have often been misinterpreted, as discussed previously (7), and it would be pertinent to reiterate the major points.

In presenting the results of our experiments, we adopted the conservative view that our initial result of a resting membrane potential less negative than -40 mV might be due to the simple fact that the isolation and purification procedures had resulted in a damaged cell sarcolemma, resulting in abnormal membrane conductances. We considered this hypothesis unlikely from the evidence of our other studies on the ultrastructure and biochemical responses of the cells (6), together with the observations that the cells were quiescent before impalement and displayed an average resting potential of almost -80 mV a few milliseconds after microelectrode insertion in solution containing 0.5 mM calcium chloride (24). The question then arose as to how those negative potentials could be maintained and we investigated a number of experimental interventions which were designed on the assumption that the true resting potentials of the cells were recorded when minimal damage occurred on microelectrode impalement, accompanied by a good sealing of the membrane around the electrode. The most convenient manoeuvre was to insert the microelectrode when extracellular calcium was 5.5 or 10.5 mM (thus calcium-tolerant cells were essential), when the myocyte then hyperpolarized to about -85 mV after about 30 seconds, a stable potential which was maintained when extracellular calcium was reduced to 2.5 or 0.5 mM (24).

Once stable recording conditions are achieved, measurements of cell input resistance, cell capacitance and membrane time-constant can be made. Healthy rat ventricular myocytes have a mean input resistance of about 40 MΩ, capacitance of the order 200 pF and membrane time constant of approximately 8 msec when measured with a glass microelectrode (24). We estimated from these data that specific membrane capacitance (C_m) was about 2.5 μF cm^{-2} and a specific membrane resistance (R_m) of 2–3 kΩ/cm^{-2}, if we used the best available estimate of cell surface area (8000 μm^2) for rat ventricular cells. These figures have often been used by others in the literature to support contentions that our myocytes were 'abnormal', while conveniently ignoring the fact that a membrane time constant of the order 8 msec is now accepted as the norm for rat ventricular cells isolated by many laboratories using a variety of dissociation techniques. In fact, adopting a membrane surface area of about 14000 μm^2 per cell, based on more recent morphological measurements on rat ventricle (27), yields estimates of 1.4 μF cm^{-2} for C_m and approximately 6 kΩ cm^2 for R_m. Alternatively, by assuming 1 μF cm^{-2} for C_m and an average cell capacitance of 200 pF, then R_m is 8 kΩ/cm^2. These values are quite comparable to measurements reported for multicellular preparations (30) and can be taken as standard for ventricular cells isolated from mammalian myocardium. It should be noted that whereas impalement in high calcium solutions is a useful manoeuvre for rat ventricular cells, resting membrane potentials of the order -80 mV can be obtained in guinea-pig ventricular cells even when microelectrode insertion is carried out in media containing 1 or 2.5 mM calcium (20).

Rapid inward Na current

The rapid upstroke of the cardiac action potential is generated by an inflow of sodium (Na) ions, but the direct measurement of this current has been considered almost impossible in cardiac tissue, due to the fact that Na current has the largest conductance and most rapid kinetics of any current in the heart. It is not an overstatement that it was only the advent of the single cardiac cell and the suction electrode that has enabled reliable analysis of rapid inward Na current (I_{Na}) to be achieved. Lee et al. (12) demonstrated the feasibility of voltage clamping single myocytes with one suction pipette, but a limitation of this approach applied to small cells having large Na currents is the voltage drop across the tip resistance of the pipette. A similar limitation is present in similar experiments reported by others (1, 28, 29).

More marked improvements in the measurement of I_{Na} was achieved with a two suction-electrode voltage clamp applied to isolated rat ventricular myocytes (2–4). Peak I_{Na} at 20–22 °C was 70–140 nA at a clamp potential of between -30 and -20 mV, with a threshold at -70 to

-60 mV. It is still the case that no reliable measurements have been made of I_{Na} in single ventricular cells at 37 °C, despite the remarkable advances that have been made in recording techniques. Translating peak I_{Na} into a peak current density requires a knowledge of the relevant area across which I_{Na} is flowing. Assuming a mean cell surface area of 14 000 μm^2, based on the evidence discussed above, then this gives a peak current density in the range 0.5–1.0 mA cm^{-2}. It is probably far better to express all currents in terms of preparation capacitance, with the proviso that this does not imply that current is uniformly flowing through this equivalent membrane area, so that I_{Na} is of the order 350–700 pA pF^{-1}.

In the double suction-electrode voltage clamp experiments great precaution was taken to minimize contaminating currents, so that maximum Na conductance ($\bar{g}_{Na}$) could be calculated from the appropriate current-voltage curves. In isolated rat ventricular cells $\bar{g}_{Na}$ is of the order 2 μS/cell (4), giving an estimate for $\bar{g}_{Na}$ as 10 nS/pF cell capacitance, in good agreement with estimates obtained from single-channel conductance measurements in neonatal heart cells (5). The ability to measure I_{Na} with a good degree of accuracy in single cells has not only provided experimental data to compare with the copious literature using V_{max} as an estimate of Na conductance, but has provided an essential experimental tool for the investigation of many pharmacological interventions which are thought to interact with Na channels, of which local anaesthetics are of particular interest and importance.

Second inward calcium current

While experiments on I_{Na} were a technical tour de force and the results indicated a more complex Na channel than found in nervous tissue, for example, work on Ca currents in single cells have produced quite revolutionary results. Our own experiments (16, 17, 25) on rat myocytes have shown that the second inward current (I_{si}') is much larger in amplitude and more rapid in kinetics than that found in multicellular preparations. Independent work on guinea-pig (13–15) and bovine (11) ventricular myocytes has reached similar conclusions. We also concluded that calcium was the major ion carrying I_{si}, and that the current could be abolished by organic and inorganic blockers. In addition, sodium was not a major charge carrier in our experiments and the time-to-peak and rate of apparent inactivation of I_{si}, together with peak current amplitude, were markedly dependent upon temperature (17).

The new characteristics of the 'second', rather than the 'slow', inward current have important implications for our understanding of many cardiac mechanisms. It is not clear how I_{si} is related to developed tension (10, 17, 18, 21) and results indicate that there must be at least two inward currents associated with the plateau phase of the cardiac action potential (16, 19, 20). Further, accurate analyses of I_{si} has resulted in a new insight into calcium channel selectivity inferred from whole-cell current records (11, 14, 17) and the mode of action of the therapeutically important 'calcium antagonist' class of drugs (15). There is no doubt that further studies of cardiac calcium channels will provide fundamental information on the initiation of the heartbeat.

Conclusions

The availability of single heart cells has revolutionized experiments in cardiac electrophysiology. The brief summary given here does not do justice to the research already carried out on sodium and calcium currents, and lack of space has precluded consideration of potassiuim currents and the wealth of information now being published using patch-clamp techniques. It is clear that the isolated cardiac cell, from whatever location in the whole organ, will be an essential preparation in the foreseeable future for many studies concerned with the electrical activity of the heart.

References

1. Bodwei R, Hering S, Lemke B, Rosenshtraukh LV, Undrovinas AI, Wollenberger A (1982): Characterization of the fast sodium current in isolated rat myocardial cells: simulation of the clamped membrane potential. J Physiol Lond 325: 301–315
2. Brown AM, Lee KS, Powell T (1980): Reactivation of the sodium conductance in single heart muscle cells. J Physiol Lond 301: 78–79P
3. Brown AM, Lee KS, Powell T (1981): Voltage clamp and internal perfusion of single rat heart muscle cells. J Physiol Lond 318: 455–477
4. Brown AM, Lee KS, Powell T (1981): Sodium current in single heart muscle cells. J Physiol Lond 318: 479–500
5. Cachelin AB, De Payer JE, Kokubun S, Reuter H (1983): Sodium channels in cultured cardiac cells. J Physiol Lond 340: 389–401
6. Dow JW, Harding NGL, Powell T (1981): Isolated cardiac myocytes: I. Preparation of adult myocytes and their homology with the intact tissue. Cardiovasc Res 15: 483–514
7. Dow JW, Harding NGL, Powell T (1981): Isolated cardiac myocytes: II. Functional aspects of mature cells. Cardiovasc, Res 15: 549–579
8. Fabiato A, Fabiato F, Sonnenblick EH (1971): Propriétés mechaniques et electriques de cellules isolées de rat adulte. J Physiol Paris 63: 47
9. Fabiato A, Fabiato F (1972): Excitation-contraction coupling of isolated cardiac fibers with disrupted or closed sarcolemmas. Circulation Res 31: 293–307
10. Isenberg G (1982): Ca entry and contraction as studied in isolated bovine ventricular myocytes. Z Naturf 37c: 502–512
11. Isenberg G, Klockner U (1982): Calcium currents of isolated bovine ventricular myocytes are fast and of large amplitude. Pflügers Arch 395: 30–41
12. Lee KS, Weeks TA, Kao RL, Akaike N, Brown AM (1979): Sodium current in single heart muscle cells. Nature Lond 278: 269–271
13. Lee KS, Lee EW, Tsien RW (1981): Slow inward current carried by Ca or Ba in single isolated heart cells. Biophys J 33: 143a
14. Lee KS, Tsien RW (1982): Reversal of current through calcium channels in dialysed single heart cells. Nature Lond 297: 498–501
15. Lee KS, Tsien RW (1983): Mechanism of calcium channel blockade by verapamil, D 600, diltiazem and nitrendipine in single dialysed heart cells. Nature Lond 302: 790–794
16. Mitchell MR, Powell T, Terrar DA, Twist VW (1983): Contribution of the second inward current to action potential time course in isolated rat ventricular myocytes. J Physiol Lond 334: 57–58P
17. Mitchell MR, Powell T, Terrar DA, Twist VW (1983): Characteristics of the second inward current in cells isolated from rat ventricular muscle. Proc R Soc Lond B 219: 447–469
18. Mitchell MR, Powell T, Terrar DA, Twist VW (1983): The time course of contraction in cells isolated from rat ventricular tissue. J Physiol Lond 345: 26P
19. Mitchell MR, Powell T, Terrar DA, Twist VW (1984): Strontium, nifedipine and 4-aminopyridine modify the time course of the action potential in cells from rat ventricular muscle. Br J Pharmac 81: 551–556
20. Mitchell MR, Powell T, Terrar DA, Twist VW (1984): The effects of ryanodine, EGTA and low-sodium on action potentials in rat and guinea-pig ventricular myocytes: evidence for two inward currents during the plateau. Br J Pharmac 81: 543–550
21. Mitchell MR, Powell T, Terrar DA, Twist VW (1984): Membrane potential and contraction in voltage-clamped cells from rat and guinea-pig ventricular muscle. J Physiol Lond 346: 77P
22. Powell T, Terrar DA, Twist VW (1978): Membrane potentials in muscle cells isolated from adult rat myocardium. J Physiol Lond 282: 23–24P
23. Powell T, Terrar DA, Twist VW (1978): Electrical activity in superfused cells isolated from adult rat ventricular myocardium. J Physiol Lond 284: 148P
24. Powell T, Terrar DA, Twist VW (1980): Electrical properties of individual cells isolated from adult rat ventricular myocardium. J Physiol Lond 302: 131–153
25. Powell T, Terrar DA, Twist VW (1981): The effect of noradrenaline on slow inward current in rat ventricular myocytes. J Physiol Lond 319: 82P

26. Rieser G, Sabbadini R Paolini P, Fry M, Inesi G (1979): Sarcomere motion in isolated cardiac cells. Am J Physiol 236: C70–77
27. Severs NJ, Slade AM, Powell T, Twist VW, Warren RL (1982): Correlation of ultrastructure and function in calcium-tolerant myocytes isolated from the adult rat heart. J Ultrastruct Res 81: 222–239
28. Undrovinas AI, Yushmanova AV, Hering S, Rosenshtraukh LV (1979): Use of the voltage clamp method in the single mammalian cardiac cells for ionic current measurement. Physiol J USSR 66: 602–606
29. Undrovinas AI, Yushmanova AV, Hering S, Rosenshtraukh LV (1980): Voltage clamp method on single cardiac cells from adult rat heart. Experimentia 36: 572–573
30. Weidmann S (1970): Electrical constants of trabecular muscle from mammalian heart. J Physiol Lond 210: 1041–1054

Author's address:

Dr. T. Powell, Department of Medical Physics and Institute of Nuclear Medicine, The Middlesex Hospital Medical School, London W1P 6 DB (U.K.)

The effect of cyanide on the K-current in guinea-pig ventricular myocytes

G. Van der Heyden, J. Vereecke and E. Carmeliet

Laboratorium voor Fysiologie, Katholieke Universiteit Leuven, Leuven (Belgium)

Summary

The mechanism of the shortening of the cardiac action potential by cyanide was studied in guinea-pig ventricular myocytes using a two micro-electrode voltage clamp technique. It is shown that the shortening can be counteracted by glucose and is due to a marked increase in K conductance.

Key words: heart, single cell, voltage clamp, metabolic inhibition, K-current

Introduction

Metabolic inhibition causes a marked shortening of the cardiac action potential in different preparations (2), mainly by decreasing the duration of the plateau phase of the action potential. Until recently the problem of the shortening of the action potential during metabolic blockade could not be solved satisfactorily because cardiac cells are known to uncouple under these conditions (12, 3), i.e. tissue loses its syncytial nature by an increase in the intercellular coupling resistance. The lack of electrical coupling between cells under conditions of metabolic inhibition makes voltage clamp analysis of the effect of metabolism on the ionic currents in multicellular preparations unreliable. In this work the mechanism of shortening of the ventricular action potential by metabolic inhibition was investigated in single ventricular myocytes using a voltage clamp technique. The use of a single myocyte obviates cellular uncoupling as a problem which makes voltage clamp experiments unsuitable for the study of metabolic inhibition in multicellular preparations.

Methods

Single myocytes were isolated from guinea-pig ventricular myocardium following a procedure similar to that described by Isenberg and Klöckner (6, 7). Cells were stored in "KB-medium" (7) at room temperature or in the refrigerator.

Small quantities of cells were transferred to the perfusion chamber (volume 1 ml), which was placed on the stage of an inverted microscope (Zeiss IM 35). KB-medium was washed out and cells were continuously superfused at a rate of about 2 ml/min with Tyrode solution at 36 ± 1 °C which was pregassed with pure oxygen. The composition of the control Tyrode solution was as follows: 150 mM NaCl, 5.4 mM KCl, 1.8 mM $CaCl_2$, 1.2 mM $MgCl_2$, 5 mM Hepes, at pH 7.4.

The experimental solutions were derived from the normal Tyrode and modifications are explicitly mentioned.

Conventional glass microelectrodes were used, filled with a 3 M KCl sulution (resistance, 15 to 30 MOhm). The experimental chamber was connected to ground via an agar bridge. The preparation could be stimulated by injection of a small constant current through the voltage-recording electrode. The voltage drop which was produced across the microelectrode during current injection was compensated electronically (5). In voltage clamp experiments a second KCl-filled microelectrode was used to inject current into the cell.

Results

When the myocytes were exposed to the calcium-containing Tyrode solution (washout of the KB-medium), a number of cells developed irreversible contracture and became spherical (Ca-paradox, see 13, 4, 10).

The remaining rod-shaped cells which showed clear cross-striations and sharp edges with resting potentials between −75 and −85 mV, were used in this study. When stimulated such cells generated action potentials with an amplitude of about 120 mV, a maximum rate of depolarization around 300 V/s and a duration between 150 and 300 ms.

After a period of equilibration in control condition, the guinea-pig ventricular myocytes were perfused with a Tyrode solution containing 1 mM NaCN. Within 5 to 10 minutes a marked shortening of the action potential was seen. Eventually the action potential lost its plateau completely and was reduced to a spike with a duration of about 20 ms (Fig. 1 upper left panel) with only slight decrease in overshoot and maximum speed of depolarization. The effects on the resting potential were somewhat variable. Most often a small initial depolarization (2 to 5 mV) oc-

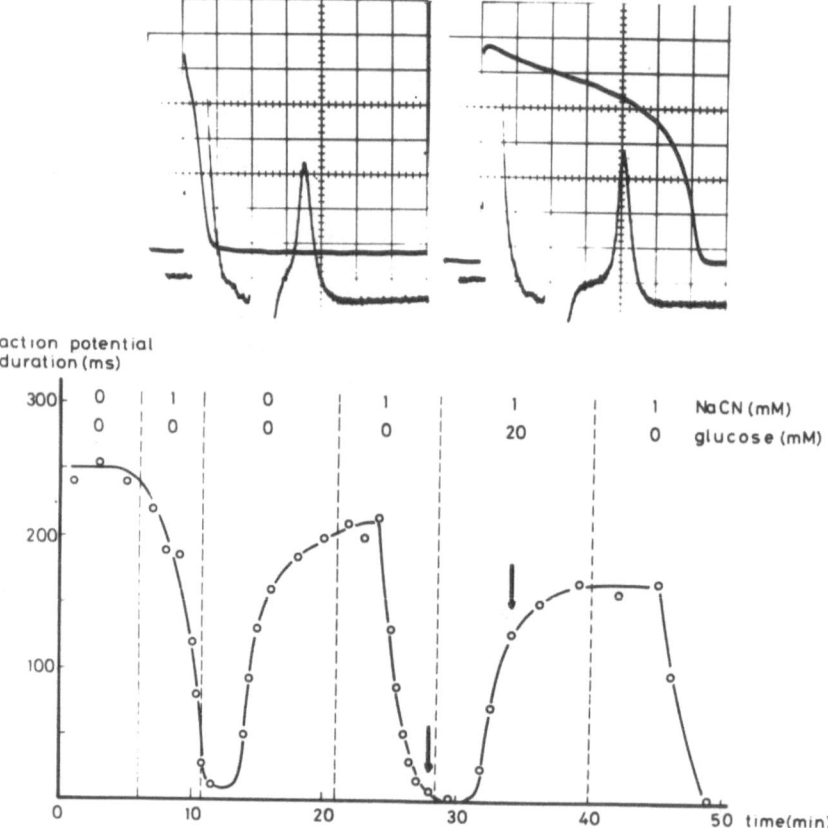

Fig. 1. Lower part: action potential duration in function of time, with or without cyanide and/or glucose in the bathing solution. The arrows correspond with the original traces in the upper part of the figure.
Upper part: action potentials and dV/dt of the upstroke, illustrating the recovery of action potential duration after adding 20 mM glucose to the CN-containing solution (arrows).
Calibrations: action potential: 20 mV and 20 ms per division dV/dt: 80 V/s and 0.5 ms per division.

curred which was eventually followed by a small hyperpolarization. The effect of cyanide was often reversible if the drug was washed within a few minutes after the action potential became spike-like in appearance (see Fig. 1 lower left panel).

These effects are comparable to the effects of hypoxia and FCCP (carbonyl cyanide 4-tri-fluoromethoxyphenylhydrazone) (unpublished results) and to the effect of DNP (2,4-dinitrophe-nol) (8) on the same preparation.

The action potential shortening induced by cyanide could be counteracted if a sufficient concentration of glucose was added to the CN-containing solution. In the experiment shown in Fig. 1, it can be seen that, after the action potential developed a spike-like appearance in 1 mM NaCN, the action potential recovered its control duration within 10 min of the addition of 20 mM glucose to the NaCN-containing solution. This experiment demonstrates that the effect of CN$^-$ is not a direct membrane effect but that it is mediated by cellular metabolism.

In order to study the ionic currents in the myocytes we used the following pulse protocol. The membrane was clamped for 500 ms at -80 mV; then a 500 ms prepulse to -45 mV was used to inactivate the fast Na current. Thereafter a test pulse to about $+10$ mV was applied for 300 ms. Finally, the clamp was released and membrane potential returned to its resting value. This voltage clamp sequence was repeated each 30 s in order to provide information on the time course of membrane currents at three different potentials during the development of CN-effect. Membrane current was measured after 300 ms at each of the three potential levels. The control isochronic current-voltage relation is indicated by the open circles in Fig. 2.

The main effect on the net membrane current of the addition of 1 mM NaCN to the external solution was a gradual but prominent increase in the time-independent outward current, which resulted in a very pronounced increase in slope of the current-voltage relation. In the experiment illustrated in Fig. 2, the slope conductance (measured at the resting potential 4 min after the onset of the CN-induced effect) is approximately 7 times higher than in control condition.

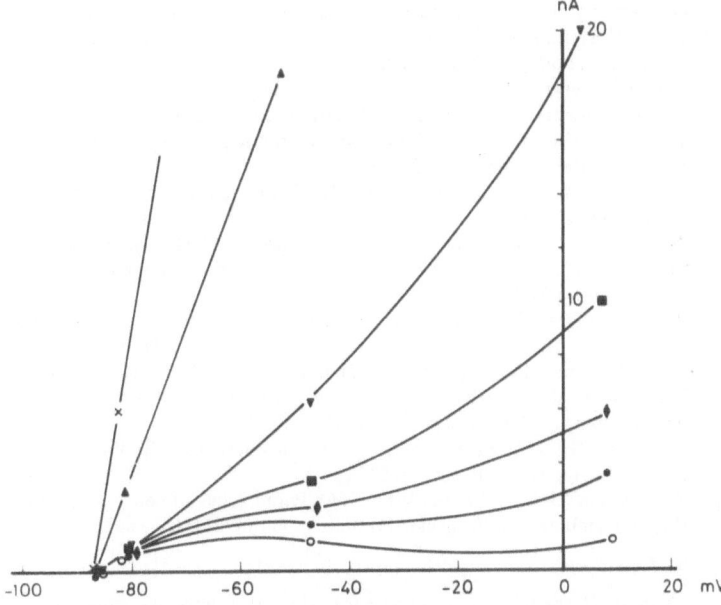

Fig. 2. Isochronic current-voltage relationships (current measured after 300 ms) in control (o) and in the presence of 1 mM NaCN (●, ◆, ■, ▼, ▲, x, with intervals of 1 min, 30 s, 30 s, 30 s, and 30 s respectively).

The cyanide-induced current reversed at a potential between -85 and -90 mV. Such a reversal potential close to E_K suggests that the largest part of the current is carried by K ions. Our findings therefore suggest that the shortening of the action potential by NaCN is due to the activation of an outward time-independent current mainly carried by K ions.

Discussion

The shortening of the cardiac action potential in the presence of cyanide in ventricular guinea-pig myocytes appears to be caused by a pronounced increase in time-independent K conductance. This effect is reversible and can be counteracted by addition of glucose to the bathing medium. Since prolonged oxygen depletion of these myocytes also causes a similar reversible shortening of the action potential, we can conclude that the observed decrease in action potential duration is not a direct effect on the membrane, rather it is mediated by metabolic inhibition.

In a previous study (8) we investigated the nature of the shortening of the action potential by DNP and were able to show that DNP produced its effect by an increased K conductance. The DNP-induced effect was not abolished in the presence of 1 mM Ba or 20 mM Cs, agents known to block the inward rectifying K channel.

The effects of CN are similar to the effects of DNP. The increase in K conductance by metabolic inhibition may be mediated by an increase in intracellular Ca activity (8). However, a direct effect of lack of ATP is also a possible alternative since ATP-regulated K channels sensitive to hypoxia and CN have been described in patch clamp experiments in guinea-pig and rabbit myocytes (1,9,11).

References

1. Bechem M, Pott L (1984): K-channels activated by loss of intracellular ATP in guinea-pig atrial cardioballs. J Physiol 348: 50 P
2. Carmeliet E (1978) Cardiac transmembrane potentials and metabolism. Circ Res 42: 577–587
3. Dahl G, Isenberg G (1980): Decoupling of heart muscle cells: correlation with increased cytoplasmic calcium activity and with changes of nexus ultrastructure. J Membrane Biol 53: 63–75
4. Dow JW, Harding NGL, Powell T (1981): Isolated cardiac myocytes. I. Preparation of adult myocytes and their homology with the intact tissue. Cardiovasc Res 15: 483–514
5. Dreyer F, Peper K (1974): Iontophoretic application of acetylcholine: advantages of high resistance micropipettes in connection with an electronic current pump. Pflügers Arch 348: 263–272
6. Isenberg G, Klockner U (1980): Glycocalix is not required for slow inward calcium current in isolated rat heart myocytes. Nature 284: 358–360
7. Isenberg G, Klockner U (1982): Calcium tolerant ventricular myocytes prepared by preincubation in a "KB-medium". Pflügers Arch 395: 6–18
8. Isenberg G, Vereecke J, Van der Heyden G, Carmeliet E (1983): The shortening of the action potential by DNP in guinea-pig ventricular myocytes is mediated by an increase of a time-independent K conductance. Pflügers Arch 397: 251–259
9. Noma A (1983): ATP-regulated K^+ channels in cardiac muscle. Nature 305: 147–148
10. Russo MA, Cittadini A, Dani AM, Inesi G, Terranova T (1981): An ultrastructural study of calcium induced degenerative changes in dissociated heart cells. J Mol Cell Cardiol 13: 265–279
11. Trube G, Hescheler J (1983): Potassium channels in isolated patches of cardiac cell membrane. Naunyn Schmiedeberg Arch Pharmacol 322: R 64, abstr. 255
12. Wojtczak J (1979): Contractures and increase in internal longitudinal resistance of cow ventricular muscle induced by hypoxia. Circ Res 44: 88–95
13. Zimmerman ANE, Hulsmann WE (1966): Paradoxical influence of calcium ions on the permeability of the cell membranes of the isolated rat heart. Nature 211: 646–647

Authors' address:

G. Van der Heyden, Laboratorium voor Fysiologie, Campus Gasthuisberg, B-3000 Leuven (Belgium)
Address for proofs:
J. Vereecke, Lab. voor Fysiologie, Campus Gasthuisberg, B-3000 Leuven (Belgium)

Measurements of single-channel currents in the membrane of isolated cells: ATP-dependence of K+-channels

G. Trube

Max-Planck-Institut für biophysikalische Chemie, Göttingen (F.R.G.)

Summary

Single-channel currents in ventricular cells of guinea-pig hearts were recorded by the patch-clamp technique. An inwardly rectifying K+-current was found in cell-attached membrane patches. Patches could be isolated from the cell exposing the cytoplasmic face of the membrane directly to the bathing solution. After isolation, the same current as in cell-attached patches was seen if the bath contained 4 mM ATP. Without ATP, this current disappeared and another channel of larger conductance and different kinetics was activated. Currents through the latter channel were also seen in cell-attached patches after poisoning the cells by DNP. It is suggested that the ATP-dependence of the observed membrane channels mediates the increase of potassium conductance after metabolic inhibition.

Key words: heart ventricle, single cell, potassium current, single channel, ATP

Introduction

Enzymatically isolated heart cells are suited for measuring single-channel currents by the patch-clamp method. In this technique (1), a glass pipette (tip diameter ~ 2 μm) is pressed against the cell surface and a seal of $10–100$ GΩ resistance is formed between the glass and the membrane. A clean membrane surface, e.g. that obtained during the enzymatic separation of cells, is a prerequisite for this "gigaseal". The electrical isolation of the small membrane patch covered by the pipette allows the resolution of the current fluctuations by the opening and closing of single ion channels. The amplitude of the current through a single channel can be measured directly and more detailed information about the kinetics of the ion channels is obtained. Various types of channels can be discriminated reliably, whereas a similar separation is hardly possible by voltage-clamp experiments on whole cells or multicellular preparations. If the pipette is drawn away from the cell, the patch is physically isolated and its cytoplasmic face is directly exposed to the bath solution ("inside-out" patch). Thus, the influence of substances on the intracellular side of the membrane can be easily tested.

Using these techniques, two types of potassium channels and their dependence on intracellular ATP were studied in ventricular cells of guinea-pig hearts (5, 6, 7).

Methods

Single cells were isolated following the method described in (2). During the experiments the cells were bathed in modified Tyrode solution containing (in mM): NaCl 140, KCl 10.8, CaCl$_2$ 1.8, MgCl$_2$ 1, glucose 10, HEPES 5 (pH = 7.4). The temperature was 21 ± 2 °C. Pipettes for current recording were filled with: KCl 145, CaCl$_2$ 1.8, MgCl$_2$ 1, HEPES 5 (pH = 7.4). The high concentration of K+ in this solution was used to increase the amplitude of the K+-currents and, thus, to improve the signal-to-noise ratio. When patches were isolated from the cells the bath was filled by a solution similar to the intracellular solution: Na$_2$ ATP 4, KCl 133, MgCl$_2$ 7.1, EGTA 2, CaCl$_2$ 0.5, HEPES 5 (pH = 7.15, pCa = 7). Alternatively, ATP was omitted from this solution substituting it by 8 mM NaCl. For details of the methods see (5, 7).

Results and discussion

When pipettes were sealed to the membrane of intact cells, step-like current fluctuations as shown by Fig. 1A were frequently observed at negative potentials. The steps are caused by the opening and closing of single ion channels. Their amplitude, i.e. the single-channel current, is linearly related to the membrane potential (slope conductance 27 pS). Close to 0 mV current fluctuations cannot be seen, since the concentration of K^+ in the pipette was approximately equal to the intracellular concentration of K^+. For smaller amounts of K^+, the zero-current potential is shifted to negative values as predicted by the Nernst-relation for a K^+-current (5). The channels rectify strongly in an inward direction, because outward currents at potentials more positive than the potassium equilibrium potential E_K cannot be observed (e.g. Fig. 1A, 40 mV). On a poster during the symposium it was shown that the rectification is due to a gate closing the channels rapidly if the potential approaches E_K or becomes more positive (8). Since the channels have various properties of the inwardly rectifying K^+ conductance of the whole cell membrane, it is suggested that they mediate the membrane current usually termed I_{K1} (5).

When the solution in the bath was replaced by the "intracellular solution" (see Methods) and the pipette was withdrawn from the cell forming an isolated "inside-out" patch (1), the kinetics of the current fluctuations were slightly modified (7). Stronger changes occurred when the ATP was removed from bathing solution: the currents observed until now disappeared and a new single-channel current of larger amplitude (conductance 80 pS) was activated (Fig. 2). This current could be blocked again by ATP (4), although the current observed initially did not reappear (Fig. 2, bottom).

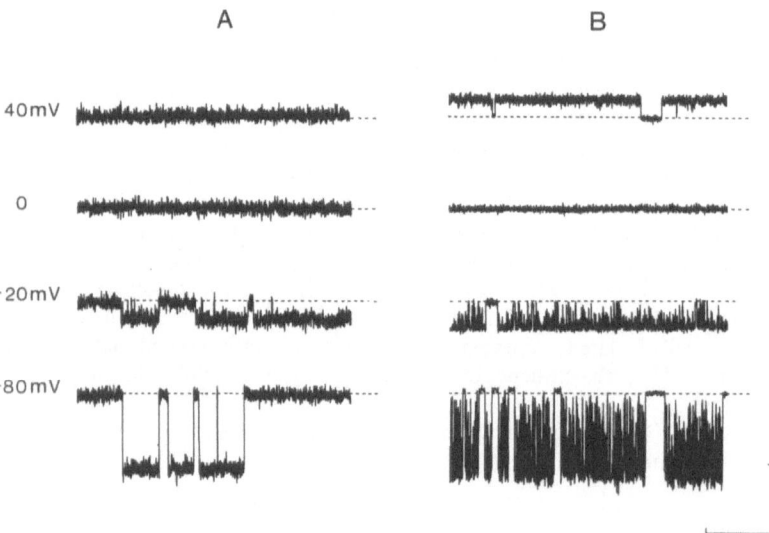

Fig. 1 A. Records of K^+ single-channel currents in a cell-attached patch at different membrane potentials given on the left.
B. Currents activated by the lack of ATP in an inside-out patch.
Dashed lines indicate the levels of the traces when the channels were closed. Inward currents (-20 mV and -80 mV) are plotted downward and outward currents (B, 40 mV) upward. At negative potentials, the channel in B frequently performs transitions to a closed state of very short mean lifetime (~ 1 ms). These steps are only partially resolved and often do not reach the dashed baseline. Calibrations: time:0.5 s, amplitude: 2 pA for A and 4 pA for B.

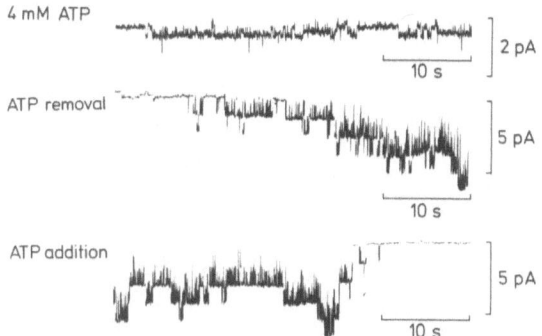

Fig. 2. Effect of ATP on the single-channel currents in an inside-out patch. Membrane potential: $-20\,\text{mV}$. Top trace: in the presence of ATP the same type of single-channel currents as in Fig. 1A is seen. Middle: removal of ATP activates the channels seen in Fig. 1B. Bottom: re-application of ATP blocks these channels. In contrast to Fig. 1 superpositions of several current steps are seen, because several channels were present in the patch.

The current activated by the lack of ATP is also carried by K^+ (7). Thus, the heart cell membrane contains two types of K^+-channels with an opposite dependence on intracellular ATP. The voltage-dependence of the channel activated by the lack of ATP is illustrated by Fig. 1B. Probably the most important difference in comparison to the other channel (Fig. 1A) is the existence of measurable outward currents.

The single-channel currents of Fig. 1B are never observed in cell-attached patches on intact cells under normal conditions. However, they can be seen after poisoning the cells by KCN or DNP (4, 7). Therefore, the activation of these channels may be the reason for the increased potassium conductance of the cell membrane during metabolic inhibition or hypoxia (3).

References

1. Hamill OP, Marty A, Neher E, Sakmann B, Sigworth FJ (1981): Improved patch-clamp techniques for high-resolution current recording from cells and cell-free membrane patches. Pflügers Arch 391: 85–110
2. Isenberg G, Klockner U (1982): Calcium tolerant ventricular myocytes prepared by preincubation in a "KB medium". Pflügers Arch 395: 6–18
3. Isenberg GF, Vereecke J, van der Heyden G, Carmeliet E (1983): The shortening of the action potential by DNP in guinea-pig ventricular myocytes is mediated by an increase of a time-independent K conductance. Pflügers Arch 397: 251–259
4. Noma A (1983: ATP-regulated K^+ channels in cardiac muscle. Nature 305: 147–148
5. Sakmann B, Trube G (1984): Conductance properties of single inwardly rectifying potassium channels in ventricular cells from guinea-pig heart. J Physiol 347: 641–657
6. Sakmann B, Trube G (1984): Voltage-dependent inactivation of inward-rectifying single-channel currents in the guinea-pig heart cell membrane. J Physiol 347: 659–683
7. Trube G, Hescheler J (1984): Inward rectifying channels in isolated patches of the heart cell membrane: ATP-dependence and comparison with cell-attached patches. Pflügers Arch 401: 178–184
8. Trube G, Hescheler J (1984): Inward rectification of single potassium channels. Basic Res Cardiol: this issue

Author's address:

Dr. G. Trube, Max-Planck-Institut für biophysikalische Chemie, Postfach 28 41, D-3400 Göttingen (F.R.G.)

Evidence for Ca-mediated inactivation of I_{Ca} in dialysed guinea-pig atrial cardioballs*)

M. Bechem and L. Pott

Institut für Zellphysiologie der Ruhr-Universität Bochum, Bochum (F.R.G.)

Summary

Ca current (I_{Ca}) was studied in cultured myocytes from right atria of adult guinea-pigs (cardioballs) by means of single low-resistance patch pipettes.

K currents were blocked by dialysis of the cells with solutions containing Cs^+ as main cation and extracellular TEA (70 mM).

Under this condition membrane currents elicited by voltage clamp pulses of 200 ms in duration from -45 mV (holding potential) are net inward for depolarizations up to $+55$ mV without detectable contamination by outward current components. The peak inward current (I_{Ca}) has a maximum between $+5$ and $+10$ mV and reverses around $+60$ mV.

Measurements of I_{Ca} tail currents obtained after clamp pulses of increasing duration to more and more positive membrane potentials suggest that I_{Ca} inactivation is not genuinely voltage-dependent.

Key words: cardioballs, cell-dialysis, Ca current, I_{Ca} inactivation, tail currents.

Introduction

Intracellular perfusion or dialysis of excitable cells with solutions of defined compositions provides a powerful method for studying properties of membrane currents with regard to kinetics as well as regulation by neurotransmitters, drugs and possible cytoplasmatic factors (e. g. 6, 8, 7). Small spherical cells can be effectively dialysed under simultaneous voltage clamp by means of a single low resistance patch clamp pipette (3). In the present study we have applied this technique to cardioballs obtained by culturing isolated atrial myocytes from hearts of adult guinea-pigs (2). Under certain experimental conditions Ca currents can be recorded from these cells with minimal interference from other membrane current components.

Methods

Experiments were performed on guinea-pig atrial cardioballs using the "tight seal whole-cell recording" technique (10). The method of dispersion and culturing of these cells has been described previously (2). Patch clamp pipettes were made from pyrex glass and were filled with a solution of the following composition (mM): CsCl, 110; EGTA, 20; $MgCl_2$, 5; Na_2ATP 1.0; c-AMP, 0.1; Hepes, 10.0, adjusted with CsOH to pH 7.3. The resistance ranged from 2–7 MΩ. 1 hour prior to an experiment the culture medium was replaced by the following solution: NaCl, 70; TEA-chloride, 70 (the 1 M TEA-stock solution was extracted with ethylether to remove any contaminant triethylamine, see (17); $MgCl_2$, 1.0; Hepes, 10.0, adjusted with NaOH to pH 7.4.

Voltage clamp experiments were performed by means of a patch clamp amplifier (List LM/EPC 7), which allowed capacitance and series resistance compensation. Signals were stored on FM tape and were analysed using a digital recorder (Datalab 4000).

All experiments were performed at ambient temperature (20–22 °C).

---

*) This work was supported by the Deutsche Forschungsgemeinschaft (SFB 114, TP A8)

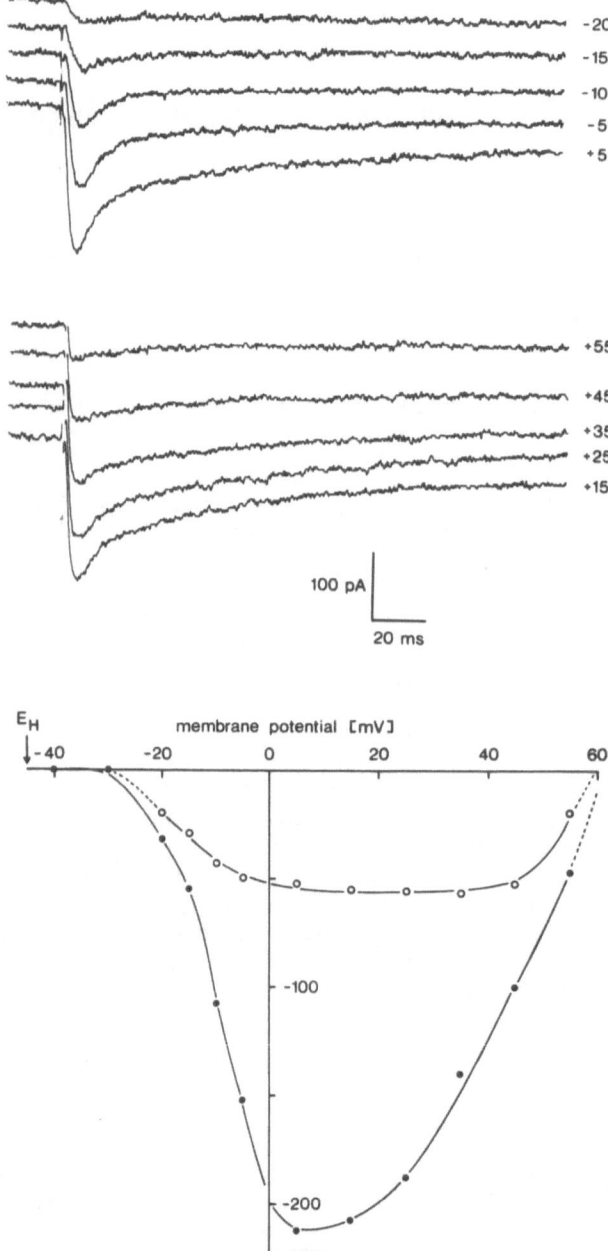

Fig. 1. Dependence of I_{Ca} on membrane potential. a) Currents elicited by voltage steps of 200 ms duration applied at 0.5 s^{-1} to various membrane potentials given on the right. Leak currents obtained from mirror-image hyperpolarizing pulses have been subtracted. b) Current-voltage relationship of peak I_{Ca} (filled circles) and late currents measured after 180 ms.

Results and discussion

Under the culture conditions used (Medium 199 with Hank's salts supplemented with 5% fetal calf serum and 25 µg/ml Gentamycin) Ca currents in atrial cardioballs are weak or even absent, which at present we cannot satisfactorily explain. However, if the filling solution of the recording pipette, which after rupture of the membrane under its tip becomes the dialysing fluid, is supplemented with cyclic AMP (c-AMP, 10^{-4} M) within ca. 5 minutes after establishing the whole cell recording configuration, stable Ca currents develop which can be recorded for 40–60 minutes.

Rundown of I_{Ca}, which has been described to occur in perfused or dialysed cells (3, 1), was slow under our experimental conditions and was usually less than 20% after 40 minutes.

The voltage dependence of I_{Ca} from a representative experiment is shown in Fig. 1. As has been described recently for isolated ventricular cells using a conventional two microelectrode voltage clamp technique (5), or frog ventricular cells using the same technique as in the present study (1), I_{Ca} activation is much faster than hitherto described for multicellular cardiac preparations (cf. 11).

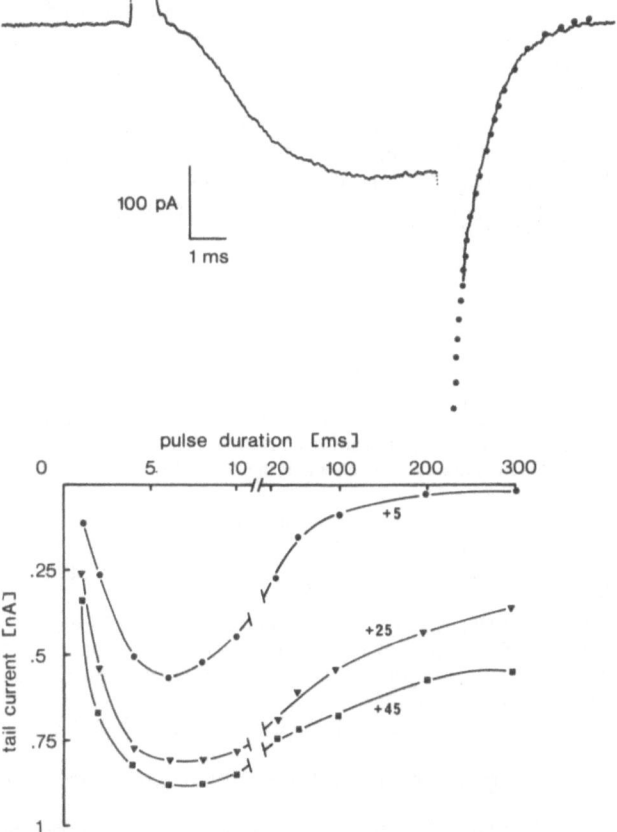

Fig. 2. Dependence of I_{Ca} tail currents on voltage step duration. a) Current trace elicited by a voltage step of 10 ms duration to +5 mV (holding potential: −45 mV) on a fast timescale. 800 µs at the beginning and the end of the pulse is blanked out due to saturation of the amplifier during capacity transients. The dotted curve represents the single exponential fit to the tail currents (see results). b) Plot of estimated tail current amplitude against duration of voltage steps to +5, +35 and +45 mV. Note the different timescales for the activating and the inactivating phase, respectively.

I_{Ca} was consistently found to be inward for depolarizations up to $+55$ mV, which applies for peak I_{Ca} as well as for the current flowing after 200 ms. At all membrane potentials tested there is a considerable fraction of I_{Ca} which does not show inactivation, ranging from about 10 to 50%. Relative to the amplitude at peak I_{Ca} the non-inactivating component is smallest between 0 and $+10$ mV, i. e. around the maximum of the current-voltage relationship. This suggests a contribution of Ca ions entering the cell to the mechanism of I_{Ca} inactivation, as has been described for a variety of excitable cells (e. g. (16), for review see (15)). Evidence for such a Ca-dependent inactivation has recently also been presented for cardiac muscle (9, 13, 12). In order to study further the time and voltage dependence of inactivation, the current tails were analysed after termination of clamp pulses with increasing duration to various membrane potentials. A typical result is illustrated in Fig. 2. Since the step change in membrane potential evokes a capacitive transient, which lasts for ca. 800 μs in the experiment shown (Fig. 2a), the amplitude of I_{tail} was evaluated by extrapolation. A semilogarithmic plot of the current relaxation starting 800 μs after the end of the clamp pulse could be fitted by a single exponential with a time constant of 780 μs. This exponential was extrapolated backwards up to a point of time 400 μs after termination of the clamp pulse. Although this procedure yields only a rough estimation of the tail current amplitudes, the data can still be used to estimate the time course of inactivation at various membrane potentials, since the error should be about the same throughout. In Fig. 2b the estimated tail currents upon repolarization to -45 mV from 3 different membrane potentials ($+5$, $+25$ and $+45$ mV) have been plotted against pulse duration. The curves clearly show that I_{Ca} inactivation is progressively slowed with increasing depolarization, i. e. with diminution of the current during the clamp pulse. In accordance with recent studies using double pulse protocols (9, 13, 12), this result is inconsistent with a genuine voltage dependence in terms of a Hodgkin-Huxley formalism (4, 14).

References

1. Bean BP, Nowycky WC, Tsien RW (1984): β-adrenergic modulation of calcium channels in frog ventricular heart cells. Nature 307: 371–375
2. Bechem M, Pott L, Rennebaum H (1983): Atrial muscle cells from hearts of adult guinea-pigs in culture: a new preparation for cardiac cellular electrophysiology. Europ J Cell Biol 31: 366–369
3. Fenwick EM, Marty A, Neher E (1982): Sodium und calcium channels in bovine chromaffin cells. J Physiol Lond 331: 599–635
4. Hodgkin AC, Huxley AL (1952): A quantitative description of membrane current and its application to conduction and excitation in nerve. J Physiol Lond 117: 500–544
5. Isenberg G., Klöckner U (1982): Calcium currents of isolated bovine ventricular myocytes are fast and of large amplitude. Pflügers Arch 395: 30–41
4. Kostyuk PG,Krishtal OA, Pidoplichko VI (1976): Effect of internal fluoride and phosphate on membrane currents during intracellular dialysis of nerve cells. Nature 257: 691–693
7. Lee KS, Akaike N, Brown AM (1980): The suction pipette method for internal perfusion and voltage clamp in small excitable cells. J Neurosci Methods 2: 58–78
8. Lee KS, Tsien RW (1982): Reversal of current through calcium channels in dialysed single heart cells. Nature 297: 498–501
9. Marban E, Tsien RW (1981): Is the slow inward calcium current of heart muscle inactivated by calcium? Biophys J 33: 143a
10. Marty A, Neher E (1983): Tight-seal whole-cell recording. In: Sakmann B, Neher E (eds). Single channel recordings. Chapter 7, Plenum, New York, 107–1
11. McDonald TF (1982): The slow inward calcium current in the heart. Ann Rev Physiol 44: 425–434
12. Mentrard D, Vassort G, Fischmeister R (1984): Calcium-mediated inactivation of the calcium conductance in cesium-loaded frog heart cells. J Gen Physiol 83: 105–131
13. Nilius B, Henček M (1983): Current dependent slow channel inactivation in heart muscle. Gen Physiol Biophys 2: 217–4

14. Reuter H (1973): Divalent cations as charge carriers in excitable membranes. Progr Biophys Molec Biol 26: 1–43
15. Reuter H (1983): Calcium channel modulation by neurotransmitters, enzymes and drugs. Nature 301: 569–574
16. Tillotson D (1979): Inactivation of Ca conductance dependent on entry of Ca ions in molluscan neurons. Proc Natn Acad Sci USA 77: 1497–1500
17. Zucker R (1981): Teteraethylammonium contains an impurity which alkalizes cytoplasm and reduces calcium buffering in neurons. Brain Res 208: 473–478

Author's address:

Dr. Martin Bechem, Bayer Forschungszentrum, Institut für Pharmakologie, 5600 Wuppertal (F.R.G.)

Cardiac membrane currents and energetic state[1])

W. Osterrieder, G. Brum

II. Physiologisches Institut der Universität des Saarlandes, Homburg/Saar (F.R.G.)

Summary

Adrenaline, cAMP and cAMP-dependent protein kinase modulate the slow inward Ca current by the same basic mechanism, presumably a phosphorylation of membrane proteins. Protein kinase also seems to play a role in the regulation of K outward currents, but not for the transient inward current.

Key words: adrenaline, cAMP, cAMP-dependent protein kinase, cardiac myocytes, voltage clamp

Introduction

Several membrane currents, which differ in ionic specificity, dependence upon membrane potential and kinetics, determine the shape of the cardiac action potential. The configuration of this action potential, in contrast to that of nerves, can vary due to numerous functionally important influences, such as heart rate, innervation or energetic state of the heart. The maximum conductances of some of these currents appear to be subject to regulatory processes. In this contribution to the symposium on "isolation, properties and applications of adult heart cells", possible mechanisms of regulation of the slow inward Ca current, the time-dependent K outward current and the (unspecific) transient inward current (1, 5) are summarized.

Methods

Ventricular myocytes from bovine hearts were prepared according to Isenberg and Klöckner (2); the method for isolating single cells from guinea-pig or cat heart has been previously described in detail (1). Composition of solutions, preparation of enzymes and experimental techniques have also been extensively described (1).

Results

Slow inward Ca current (I_{si})

Fig. 1 shows an example of intracellularly recorded action potentials from a guinea-pig ventricular cell. The administration of adrenaline caused a concentration-dependent prolongation of the action potential and a shift of the plateau to more positive levels. These effects are due to an increase in slow inward Ca current (I_{si}) (cf. (6)), mediated by an increase in intracellular cyclic AMP levels (8). Injection of CAMP into single ventricular cells produced the same effects on the action potential as extracellular adrenaline (7). The increase in I_{si} has been shown directly in voltage-clamp experiments with guinea-pig myocytes (1).

---

[1]) This work was supported by the Deutsche Forschungsgemeinschaft. Dr. Brum was a recipient of a fellowship from the German Academic Exchange Service (DAAD).

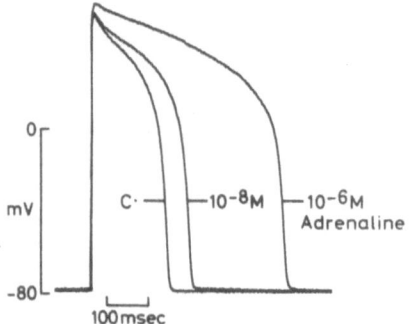

Fig. 1. Intracellularly recorded action potentials from a guinea-pig ventricular myocyte. Adrenaline (10^{-8} and 10^{-6} M) caused a concentration-dependent increase and prolongation of the action potential.

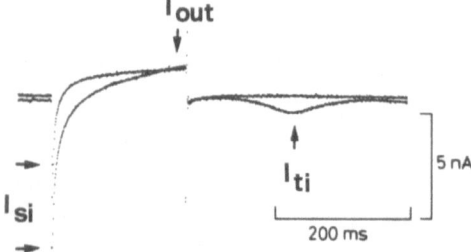

Fig. 2. The membrane potential of a cell was voltage-clamped to -55 mV and clamp steps to $+5$ mV were applied. The currents before (upper trace) and after (lower trace) an injection of the catalytic subunit of the cAMP-dependent protein kinase are superimposed. I_{si} increased markedly, I_{out} increased only slightly, and a transient inward current (I_{ti}) appeared.

Several authors have suggested that the increase in I_{si} depends on the activation of a protein kinase by cAMP, presumably by phosphorylation of a membrane protein intimately related to the Ca channel. It could indeed be demonstrated (1,5) that pressure injection of the (active) catalytic subunit of cAMP-dependent protein kinase elicited the same effects on the action potential as exposure to adrenaline or intracellular cAMP-injection. Fig. 2 shows an experiment in which the catalytic subunit was injected and membrane currents were simultaneously recorded (two-microelectrode voltage clamp). Compared with the control, protein kinase injection increased the maximum amplitude of I_{si} (indicated in the figure) by more than twofold. Analysis of the current-voltage relationship indicated that the potentials of maximum I_{si} and of current reversal were not changed.

Injection of the regulatory subunit of cAMP-dependent protein kinase decreased the Ca conductance of the cell membrane (1). These effects on the action potential were reversible, in contrast to those due to the catalytic subunit, suggesting the activation of compensatory mechanisms following injection of the regulatory subunit.

Potassium outward currents

The recording in Fig. 2 suggests that the catalytic subunit slightly increased the K outward current (I_{out}) at the end of the depolarizing test pulse (duration 200 ms, test potential $+5$ mV). This effect was clearer at more positive test potentials and longer test pulses. Current-voltage relations (Fig. 4 in (1)) before and after injection of the catalytic subunit indicated that both the time-independent and time-dependent K outward currents were increased.

Transient inward current

In many experiments, transient depolarizations accompanied by after-contractions of the cell were observed after injection of the catalytic subunit (1). These are caused by a transient inward current (I_{ti}) which differs from I_{si}. The appearance of I_{ti} after catalytic subunit injection was observed in the experiment shown in Fig. 2. These transient depolarizations often appeared with a delay after enzyme injection, disappearing after interruption of regular pulsing, and then reappearing slowly. Furthermore, they were suppressed when I_{si} was inhibited by Ca channel blocking agents (1). Thus, it is possible that I_{ti} is due to an increased intracellular Ca level resulting from the injection of the catalytic subunit.

Discussion

The major result of these studies was the demonstration of an increase of the slow inward Ca current by the catalytic subunit of cAMP-dependent protein kinase. This is strong supporting evidence for the hypothesis that phosphorylation of one or more proteins by the catalytic subunit will lead to an increase in I_{si}. It is highly probable that adrenaline, cAMP and cAMP-dependent protein kinase act through the same final step, i. e., phosphorylation of the Ca channel. It also seems likely that this mechanism does not only play a role during β-adrenergic stimulation but does so also in its absence. This is suggested by the decrease in Ca conductance observed after injection of the regulatory subunit. It is probable that the injected regulatory subunit, by binding both cAMP and the catalytic subunit, will decrease the ability of the latter to phosphorylate Ca channels. The effects of the regulatory subunit were reversible, suggesting the triggering of some compensatory mechanisms, possibly increased cAMP formation.

It seems that K channels also undergo phosphorylation. An increased cytostolic Ca concentration is unlikely to be responsible for the effect of the catalytic subunit, since an increase in outward K currents during β-stimulation has been observed even when I_{si} was completely blocked (3). The transient inward current may be explained by an increase in intracellular Ca due to enhanced Ca conductance of the cell membrane after enzyme injection. I_{ti} can also be directly induced by injection of Ca into single heart cells (4). In the present experiment, the appearance of I_{ti} presumably reflects an impaired ability of the cell to sufficiently extrude Ca by means of sarcolemmal Na/Ca exchange or Ca-ATPase mechanism.

In conclusion, the cytoplasmic concentrations of ATP and cAMP are essential for maintenance and modification of the electrical activity of the cardiac cell membrane.

Acknowledgement

We wish to thank Drs. V. Flockerzi and F. Hofmann, Pharmakologisches Institut der Universität Heidelberg, F.R.G., for the enzyme preparations.

References

1. Brum G, Flockerzi V, Hofmann F, Osterrieder W, Trautwein W (1983: Injection of catalytic subunit of cAMP-dependent protein kinase into isolated cardiac myocytes. Pflügers Arch 398: 147–154
2. Isenberg G, Klöckner U (1982): Calcium tolerant ventricular myocytes prepared by preincubation in a "KB" medium. Pflügers Arch 395: 6–18
3. Kass RS, Wiegers SE (1982): The ionic basis of concentration-related effects of noradrenaline on the action potential of calf cardiac Purkinje fibers. J Physiol 322: 541–558
4. Matsuda H, Noma A, Kurachi Y, Irisawa H (1982): Transient depolarization and spontaneous voltage fluctuations in isolated single cells from guinea-pig ventricles. Circ Res 51: 142–151

5. Osterrieder W, Brum G, Hescheler J, Trautwein W, Hofmann F, Flockerzi V (1982): Injection of sub-units of cyclic AMP-dependent protein kinase into cardiac myocytes modulates Ca^{2+} current. Nature 298: 576–578
6. Reuter H, Scholz H (1977): The regulation of the calcium conductance of cardiac muscle by adrenaline. J Physiol 264: 49–62
7. Trautwein W, Taniguchi J, Noma A (1982): The effects of intracellular cyclic nucleotides on the action potential and the acetylcholine response of isolated cardiac cells. Pflügers Arch 392: 3078–314
8. Tsien R W (1973): Adrenaline-like effects of cyclic AMP in cardiac Purkinje fibres. Nature New Biol 245: 120–122

Authors' address:

Dr. W. Osterrieder, Pharma Research Department, F. Hoffmann-La Roche & Co., Ltd. PF/1, Bau 70/426, CH-4002 Basel (Switzerland)

Simultaneous measurements of action potentials and contractions in single cultured adult and embryonic heart muscle cells[1])

B. Koidl, G. Zernig, and H. A. Tritthart

Universitäts-Institut für Medizinische Physik und Biophysik Graz (Austria)

Summary

Using both a microphotometrical technique and conventional microelectrodes, excitation and contractions have been measured simultaneously in spontaneously active and electrically driven isolated cultured heart muscle cells of embryonic chicks as well as adult rats and guinea-pigs. This experimental arrangement makes it possible to measure continuously and with high accuracy, excitation and E.C. coupling at the single cell level. The contraction time course of spontaneously active embryonic cells and electrically driven adult cells resembles contractions of macroscopic heart muscle preparations against various preloads; mostly a nearly isometric contraction form is found. The effects of various inotropic factors on excitation and contraction (beat frequency, Ca, epinephrine) were studied and it could be shown that the embryonic cells show different features of E.C. coupling. The activity of adult cells resembles very strongly that characteristically found in macroscopic preparations. This proved the feasibility of the method in studies of cardiac E.C. coupling at the single cell level.

Key words: cultured cardiac myocytes, action potential, contraction, E.C. coupling, force-frequency relationship

Introduction

In earlier publications a procedure protocol was presented which allows the measurement of transmembrane action potentials of cultured embryonic chick heart muscle cells simultaneously with their mechanical activity using microelectrodes and a microphotometrical device. It could be shown that inotropic influences (beat frequency, Ca, heart glycosides) as well as influences on the excitation process (ionic channel blockers, ionic concentrations) could thus be studied at the level of the single cell by utilizing the various advantages of this preparation (3, 4, 5).

In the present investigation this experimental system was applied for the first time to single cells from hearts of adult rats and guinea-pigs. In these cells basic features of excitation and contraction coupling could be studied and compared with findings in embryonic chick heart muscle cells.

Methods

The preparation and culture of embryonic heart muscle cells has been extensively described in earlier publications (3, 5). Cells of the hearts of adult rats and guinea-pigs were prepared using the methods of Piper et al. (6). The microelectrode technique applied as well as the microphotometrical measurement of cell pulsations have also already been documented (3, 5).

---

[1]) Supported by the Austrian Science Research Fund, grant No. 4662

Results

Using a combination of conventional microelectrode- and microphotometrical techniques action potentials and contractions were measured simultaneously from single embryonic chick heart muscle cells and of myocytes from hearts of adult rats and guinea-pigs (Figs. 1, 3b). The embryonic cells are spontaneously active and their action potentials show typical pacemaker characteristics (Fig. 1b). Some of the adult cells are also spontaneously active and beat at 37 °C with a low frequency of about 4/min; most others, however, show no spontaneous activity (cf. 1).

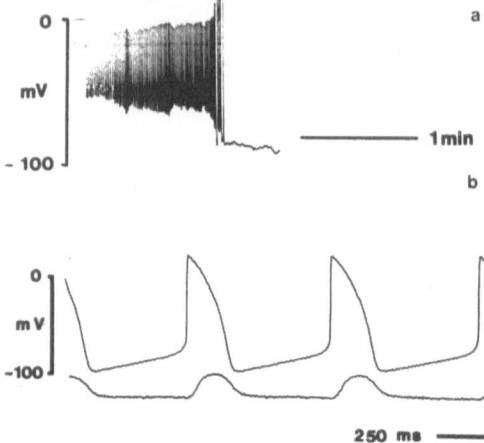

Fig. 1a. Action potentials of an adult guinea-pig heart muscle cell immediately after impalement with the microelectrode. 1b. Pacemaker action potentials (upper line) and contractions simultaneously measured in a spontaneously beating embryonic chick heart muscle cell.

In adult cells without spontaneous activity a train of spontaneous action potentials is found immediately after penetration with a microelectrode (Fig. 1a). The amplitude of these action potentials grows beat by beat until finally the repolarization again reaches the stable resting potential region and stays quiescent at about -90 mV. In the adult cells spontaneously active at a low rate, the contractions look like waves creeping over the cell. The contractions of adult cells after electrical stimulation as well as the spontaneous contractions of embryonic heart cells show time courses comparable to contractions found in macroscopic preparations against various preloads, i.e. isotonic up to isometric contraction forms.

The effects of various inotropic influences on excitation-contraction coupling were studied in cells stimulated by short current pulses at various frequencies or while the spontaneous frequency of embryonic cells was changed through long lasting intracellular current application. All cells react sensitively to changes in $[Ca]_o$ or application of cardiac glycosides. With epinephrine only the adult cells could be stimulated, i.e. the embryonic chick cells are without epinephrine sensitivity. The relationship between stimulation frequency and contraction strength is clearly different in embryonic and in adult cells and 3 examples are given in Figs. 2a, b and 3a,c. In the experiment given in Fig. 2a the spontaneous action potentials of an embryonic cell (upper row) were interrupted by hyperpolarizations lasting about 1 min. This resulted in a reduction of beat frequency and an increase in action potential amplitude. Concurrent with these changes is a reduction in mechanical resting tension and an increase in the contraction amplitude (lower row). After cessation of the current pulse the spontaneous activity reappeared, the contractions produced a positive "Treppe" (Fig. 2b) and the resting tension rose again. The action potentials seen after removal of those current pulses show a transient increase in amplitude and are followed by contractions which reflect this transient effect.

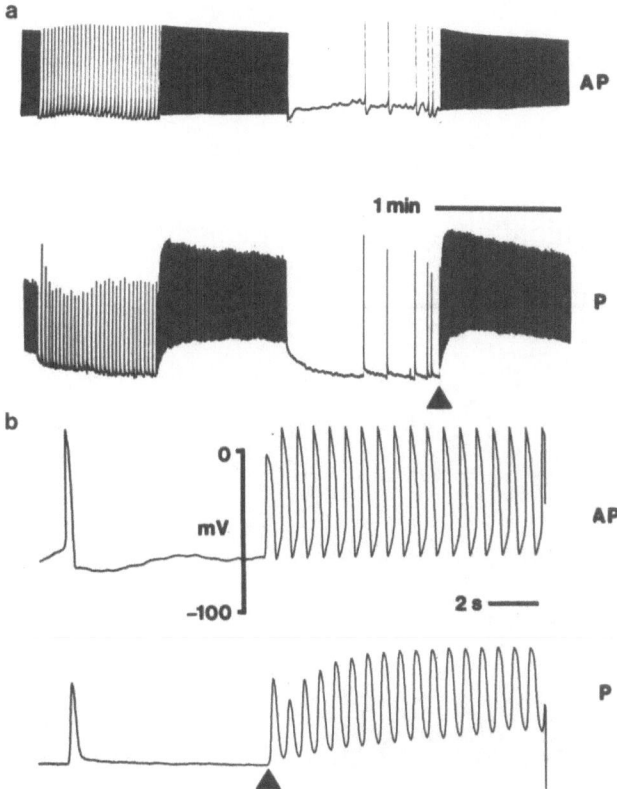

Fig. 2a. Action potentials (AP) and pulsations (P) of an embryonic chick heart cell. The sequence of events marked by an arrow is given in 2b with higher time resolution. A complete description of the protocol is given in the text.

In the adult rat cells contraction residues occur only at high stimulation frequency (4/s). Further differences are the characteristic transients after start of electrical stimulation and a negative force-frequency relationship – not unexpected for rat heart cells (2). Cells from hearts of adult guinea-pigs, on the other hand, show a positive "Treppe" (Fig. 3a).

A single action potential from the control period of the experiment given in Fig. 3a (stimulation frequency of 30/min) was compared with an action potential from the period with elevated stimulation frequency (120/min) and both are superimposed in Fig. 3b. It is evident that the elevation of the stimulation frequency causes a shortening of the action potential duration and a reduction the electromechanical latency of the single cell.

Discussion

Using a microphotometrical technique along with conventional microelectrodes (3, 4, 5), excitation and contraction of isolated cultured heart muscle cells from embryonic chick hearts, adult rats and guinea-pigs have been measured simultaneously. Whereas the embryonic cells always beat spontaneously, only few adult cells are spontaneously active (cf. 1) at a low rate. The con-

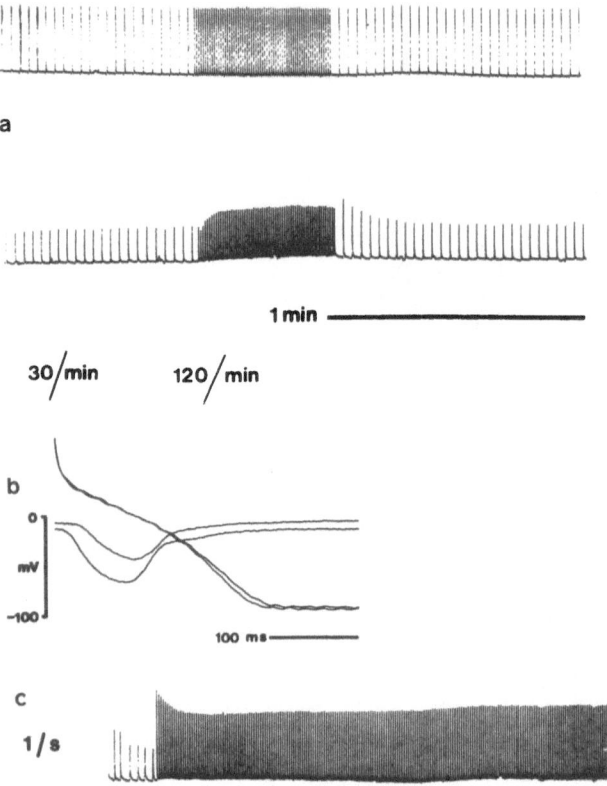

Fig. 3a. Simultaneously measured action potentials (upper line) and contractions (lower line) of an adult guinea-pig cell at different stimulation frequencies. In 3b two action potentials from this experiment at the two different stimulation rates are superimposed at a higher time resolution. 3c: contractions of an adult rat heart cell, which was spontaneously active at a low rate at the start of the registration and was then stimulated with 1/s. The negative "Treppe" characteristic for rat myocardium is evident.

tractions of the spontaneously beating embryonic and electrically driven adult cells show a time course resembling that of isotonic contractions of macroscopic heart muscle preparations under various preloads up to isometric contractions. The contraction form depends apparently on different internal mechanical factors and on different strengths of attachment of the cell to the culture glass. Comparing some inotropic influences (stimulation frequency, Ca, epinephrine, heart glycosides) it was found that the embryonic cells show less developed properties of the E.C. coupling system than the adult cells (i.e. contraction residue at low frequency, dependence of contraction strength on action potential amplitude).

These first experiments indicate that the adult cell is an adequate model for studies of the cardiac E.C. coupling process at the single cell level. The various refined methods applicable to these cells as well as numerous systemic advantages (e.g. lack of narrow intercellular clefts), emphasize the general importance of this model.

References

1. Dow JW, Harding NGL, Powell T (1981): Isolated cardiac myocytes. II. Functional aspects of mature cells. Cardiovasc Res 15: 549–579
2. Hoffmann BF, Kelly JJ (1959): Effects of rate and rhythm on contraction of rat papillary muscle. Am J Physiol 197: 1199–1204
3. Koidl B, Tritthart HA (1980): The effects of ouabain on the electrical and mechanical activities of embryonic chick heart cells. J Mol Cell Cardiol 12: 663–673
4. Koidl B, Tritthart HA (1982): D-600 blocks spontaneous discharge, excitability and contraction of cultured embryonic chick heart cells. J Mol Cell Cardiol 14: 251–257
5. Koidl B, Tritthart HA, Erkinger S (1980): Cultured embryonic chick heart cells: photometric measurement of the cell pulsation and the effects of calcium ions, electrical stimulation and temperature. J Mol Cell Cardiol 12: 165–178
6. Piper HM, Probst I, Schwarz P, Hütter FJ, Spieckermann PG (1982): Culturing of calcium stable adult cardiac myocytes. J Mol Cell Cardiol 14: 397–412

Author's address:
Dr. Bernd Koidl, Universität Graz, Institut für Medizinische Physik und Biophysik, Harrachgasse 21, A-8010 Graz (Austria)

The two components in the shortening of unloaded ventricular myocytes: Their voltage dependence

G. Isenberg, A. Beresewicz, D. Mascher and F. Valenzuela

II. Physiologisches Institut, Universität des Saarlandes, Homburg (F.R.G.)

Summary

In isolated myocytes from mammalian ventricles a fast and a slow component in the contractile response to depolarizing voltage clamp steps were identified. The potential dependence of the slow component was identical to the activation curve of i_{Ca}. The fast component, however, remained at its maximal amplitude at potentials positive to $+10$ mV (up to $+100$ mV), in which potential range i_{Ca} declined and eventually disappeared. The results suggest that the slow component may be activated by Ca^{++} entering through sarcolemmal Ca channels, whereas the fast component depends on Ca release from intracellular sites and may depend on both Ca_i and voltage.

Key words: contraction, calcium, electro-mechanical coupling, sarcomere length, isotonic contraction

Introduction

Usually, the contractility of ventricular tissue is measured as force developed under isometric conditions. Because of the difficulties of attaching a force transducer to an isolated cell, we characterize the contractile state of the unloaded cell by measuring with optical methods the change in the average sarcomere length (SL) (7). It has been shown that unloaded shortening and force follow a similar dependence on the activator calcium (a[Ca]) (3) suggesting that the sources of a(Ca) may be the same. The literature suggests several mechanisms by which a(Ca) could be increased: i) release of Ca from intracellular stores (SR, (5)), ii) transsarcolemmal Ca entry via the i_{Ca} (1, 6), iii) transsarcolemmal Ca entry via an electrogenic Na, Ca exchange (2, 12). As the transmembrane potential should influence all 3 mechanisms but in different ways, we have chosen to study the dependence of contraction on the potential of depolarizing clamp steps.

Method

Myocytes were isolated from guinea-pig or bovine ventricles as described previously (8). The superfusing Tyrode solution contained 150 mM NaCl, 5.4 mM KCl, 3.6 mM $CaCl_2$, 1.2 mM $MgCl_2$, 10 mM glucose, 5 mM Hepes/NaOH (pH 7.4). The temperature was 35 °C. The voltage clamp experiments were performed either with 2 microelectrodes (9) or with a single patch electrode (6). The contraction was monitored as edge movement with a TV camera-tape system and a photodiode array (7). The signal was calibrated in terms of the average sarcomere length (SL; SL_d for the diastolic and SL_s for the systolic SL). The maximal rate of shortening (MRS) was obtained by dividing the fastest change in sarcomere length by the 20 ms interval between consecutive semi-images of the 50 Hz TV system.

Results and discussion

Voltage dependence of the slow component of contraction (SCC)

To isolate SCC, we inactivated the fast component of contraction (FCC) by depolarizing the cells with Tyrode solutions containing 20 mM KCl (11). The cells responded to depolarizing

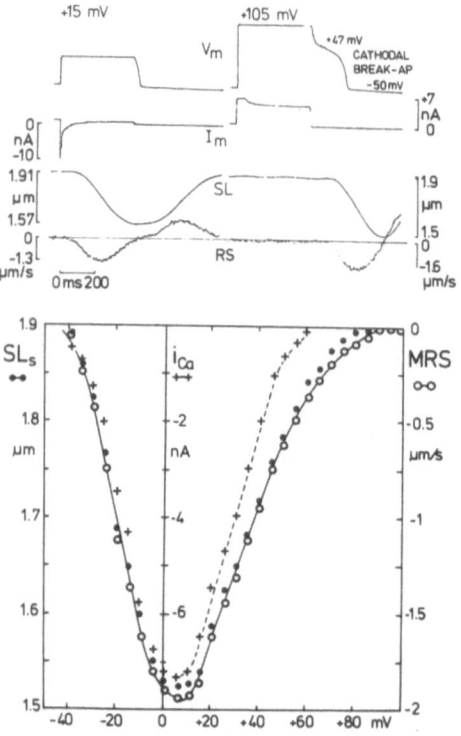

Fig. 1. Influence of the clamp step potential on membrane current and the slow component of contraction. Bovine ventricular myocyte superfused with Tyrode solution containing 20 mM KCl. 400 ms long depolarizations starting from a holding potential of −45 mV were applied at 0.1 Hz.

Upper part: membrane potential (V_m, top), net membrane current (I_m), shortening of the average sarcomere length (SL) and rate of shortening (RS). The signals SL and MRS were filtered at 10 Hz.

Note for the clamp step to + 105 mV (right): The strong depolarization does not activate the contraction, i. e. the cell remained quiescent during the depolarization. Switching off the clamp resulted in a "cathodal break AP" because the strong depolarization did activate but not inactivate the calcium conductance. Thus, this AP was generated by the inflowing Ca ions which also trigger the contraction.

Lower part: Together with the peak calcium inward current (i_{Ca}, crosses), the minimal systolic sarcomere length (SL_s, dots) and the maximal rate of shortening (MRS, circles) were plotted as a function of the clamp depolarization.

clamp steps with i_{Ca} and a contraction (Fig. 1, + 15 mV) which we consider to represent SCC because i) its onset follows the start of depolarization with a delay of 80 ms, ii) its MRS is low (− 1.3 µm/s), and iii) the shortening continues throughout the depolarization. When depolarized to + 105 mV the myocyte no longer contracts, the traces of both SL and RS show no deflection. A contraction was only evoked when the clamp was switched off and the membrane repolarized with a cathodal break AP (which is supposed to transport Ca via non-inactivated Ca-channels into the cell). The results of Fig. 1 suggest that SCC requires a(Ca) that enters via i_{Ca}. In order to quantify this requirement, we compared the influence of the clamp step potential on SL_s and MRS with that on i_{Ca} (lower part of Fig. 1). After appropriate scaling, between threshold (− 35 mV) and + 15 mV the activation curves of i_{Ca} and of SCC were nearly identical. This result suggests a causal dependence, i. e. SCC being "directly" activated by calcium entering with i_{Ca}.

At clamp potentials positive to + 20 mV, the curves for peak i_{Ca} and SCC diverge: i_{Ca} decreases more steeply to disappear at + 60 mV, whereas SL_s and MRS fall along a flatter curve which approaches + 90 mV. This "dissociation" between i_{Ca} and contractility does not argue against a "direct" activation of contraction via i_{Ca}, and it can be explained in the following 2 ways: i) We evaluated i_{Ca} by the method of the "visual estimate" which subtracts the current flowing 200 ms after start of depolarization from the surge of negative current (9). This method provides reliable data only when i_{Ca} is not superimposed on other time dependent currents, e. g. potassium currents. This "masking" of i_{Ca} by i_K prevents quantitation of i_{Ca} at strongly positive potentials. ii) Even if + 60 mV is the "i_{Ca} reversal potential" this does not mean that Ca ions flow outward. Because of their imperfect selectivity (14, 15), the Ca channels will also allow K-ions to pass. i_{Ca} due to K-transport may be outward at potentials positive to + 60 mV, but Ca ions will still enter since Ca-entry will disappear only at the calcium equilibrium potential of about + 120 mV.

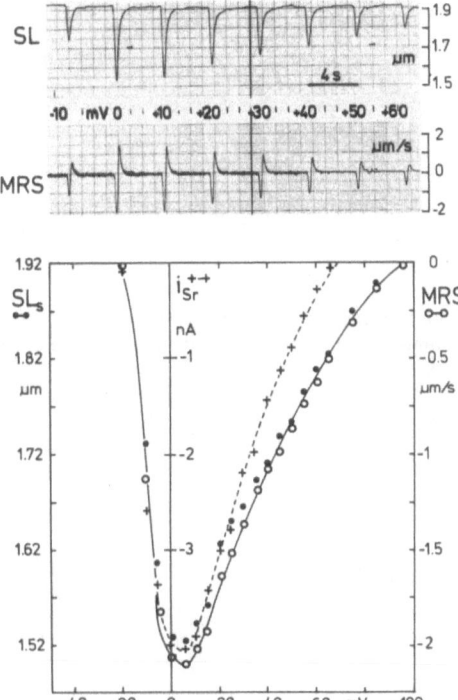

Fig. 2. Voltage dependence of the peak Sr inward current (crosses) and of the slow component of contraction as indicated by the SL_s (dots) and the MRS (circles). Guinea-pig ventricular myocyte superfused with a Tyrode solution containing no calcium but 3.6 mM strontium. The 160 ms long depolarizations started from a holding potential of -45 mV and were applied at rate of 0.25 Hz. Top: original pen recordings (filtered at 10 Hz).

A voltage dependence of SCC similar to that in Fig. 1 was recorded in 11 bovine ventricular myocytes (20 mM $[K]_o$). We also observed a similar behaviour in guinea-pig ventricular myocytes stimulated at low frequencies. The voltage dependence suggests that activation of contraction does not necessarily require the activity of an electrogenic Na, Ca exchange mechanism because clamp steps progressively more positive to the supposed reversal potential of -10 mV (12) should provoke the entry of Ca ions at a faster rate and in a greater amount, yet the contractility declines.

Another way of studying SCC selectively is to superfuse the myocyte with a Tyrode solution (5.4 mM $[K]_o$) containing 3.6 mM strontium but no calcium (10). Fig. 2 shows the contractions (top) together with curves of the peak Sr current, of the SL_s and of the MRS. As in the case of Ca currents, the curves run parallel to each other up to about $+15$ mV but diverge at more positive potentials. Our results in Sr-containing, Ca-free media again suggest that the late component of contraction is "directly" activated in this case by the Sr ions entering as a Sr current through transsarcolemmal "Ca-channels". (We may exclude a significant activation of the release of Sr from the SR because the gating of this process needs 10 times higher $[Sr]_i$ than $[Ca]_i$ [A. Fabiato, personal communication].).

The fast component of contraction

We studied FCC in isolated guinea-pig ventricular myocytes stimulated at 0.5 or 1 Hz. Fig. 3 shows the build-up of the FCC during a positive "Treppe". At steady state (Fig. 3, stS), FCC shortens the sarcomeres with a MRS of -3.4 μm/s to 1.50 μm, and this SL_s is reached within

120

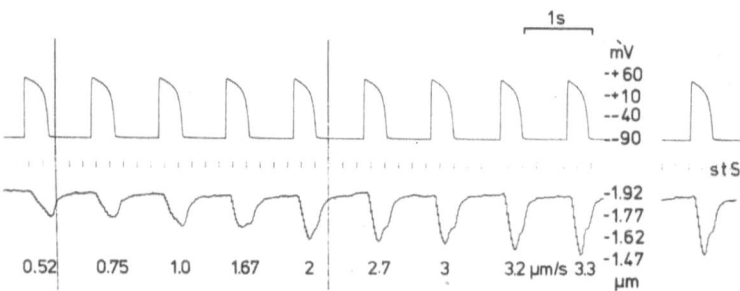

Fig. 3. The fast component of the contraction as it develops during a "positive Herztreppe". Following a 1 min rest period, the guinea-pig ventricular myocyte was stimulated at 1 Hz. MRS is labeled by numbers. stS shows the steady state with the 40th beat. Unfiltered pen record.

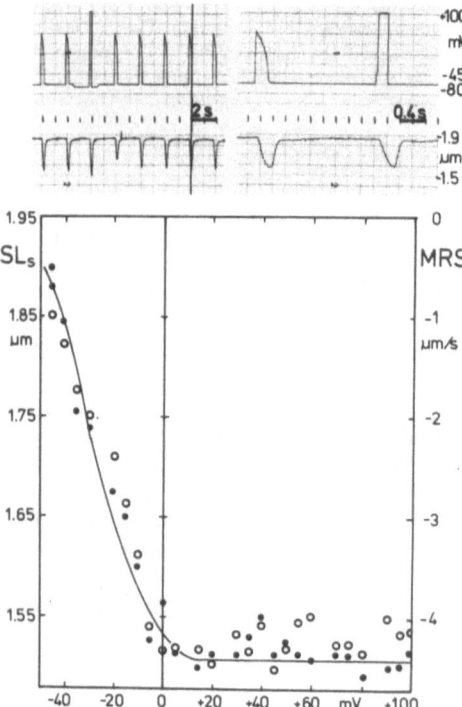

Fig. 4. Voltage dependence of the fast component of contraction. Top left: type of the protocol interposing the clamp steps between APs. Top right: shape of the contraction during an AP and a pulse to +100 mV. (The pulse started from a holding potential of -45 mV which lasted for 20 ms only, after the 160 ms long pulse the membrane was clamped back to -80 mV). Bottom: SL$_s$ (dots) and MRS (circles) as a function of the clamp step potential.

140 ms. FCC does not last as long as the plateau of the action potential, the sarcomeres relax incompletely to 1.62 μm. (We attribute this second level to the SCC).

FCC was much more labile than SCC. To analyse the voltage dependence of FCC, we stimulated the myocytes rhythmically at 0.5 or 1 Hz and interposed the voltage clamp steps between the AP_s (Fig. 4, top left). At membrane potentials between the threshold ($-40\,mV$) and $+10\,mV$, FCC followed the activation curve of i_{Ca}. At more positive potentials, however, peak i_{Ca} declined and disappeared, whereas FCC remained maximally activated (Fig. 4). The contractions, e. g. at $+100\,mV$, started 40 ms after depolarization and reached, during the clamp step, a similar SL_s to that found during the preceding AP. During the $+100\,mV$ pulse, the cell shortened with a MRS which was slightly slower (80–90%) than the MRS during the AP. The next AP evoked a slightly attenuated contraction (as might be expected if the stores were depleted) and within the following 5 beats contractility completely recovered.

If we follow conventional ideas (cf. ref. 13), we attribute the FCC to a(Ca) released from the SR. Between -40 and $+10\,mV$, the contraction shows a similar activation curve to that of the peak i_{Ca} (Fig. 4); in other words it would be consistent with the hypothesis of the "Ca release induced by $(Ca)_i$" (5). However, with this mechanism of "Ca-release induced by $(Ca)_i$" it is difficult to explain the voltage independent part of the fast component of contraction which is recorded at strongly positive potentials. If a net transsarcolemmal Ca entry is truely absent at $+100\,mV$, the maximal activation of the fast component might then be attributed to a Ca_i independent Ca release mechanism such as the "depolarization-induced Ca release" (4).

References

1. Beeler GW, Reuter H (1970): The relation between membrane potential, membrane currents and activation of contraction in ventricular myocardial fibres. J Physiol 207: 211–229
2. Chapman RA (1983): Control of cardiac contractility at the cellular level. Am J Physiol H535–H552
3. DeClerck NM, Claes VA, Brutssaert DL (1977): Force velocity relations of single cardiac muscle cells. J Gen Physiol 69: 221–241
4. Endo M (1977): Calcium release from the sarcoplasmic reticulum. Physiol Rev 57: 71–108
5. Fabiato A (1983): Calcium-induced release of calcium from the cardiac sarcoplasmic reticulum. Am J Physiol 245: C1–C14
6. Hamill OP, Marty A, Neher E, Sakmann B, Sigworth FJ (1981): Improved patch-clamp techniques for high-resolution current recording from cells and cell-free membrane patches. Pflügers Arch 391:85–100
7. Isenberg G (1982): Ca entry and contraction as studied in isolated bovine ventricular myocytes. Z Naturforsch 37c: 502–512
8. Isenberg G, Klöckner U (1982): Calcium tolerant ventricular myocytes prepared by preincubation in a "KB medium". Pflügers Arch 395: 6–18
9. Isenberg G, Klöckner U (1982): Calcium currents of isolated bovine ventricular myocytes are fast and of large amplitude. Pflüglers Arch 395: 30–41
10. King BW, Bose D (1983): Mechanism of biphasic contractions in strontium-treated ventricular muscle. Circ Res 52: 65–75
11. Mascher D, Cruz A (1980): Electrical and mechanical responses of the guinea-pig ventricular muscle in the presence of histamine. XXVIII International Congress of Physiological Sciences. Budapest, 1980
12. Mullins LJ (1981): Ion transport in heart. Raven Press, New York
13. Reiter M, Vierling W, Seibel K (1984) Where is the origin of the activator calcium in cardiac ventricular contraction? Basic Res Cardiol 79: 1–8
14. Reuter H, Scholz H (1977): A study of the ion selectivity and the kinetic properties of the calcium dependent slow inward current in mammalian cardiac muscle. J Physiol 264: 17–47
15. Tsien RW (1983): Calcium channels in excitable cell membranes. Ann Rev Physiol 45: 341–358

Author's address:

Dr. G. Isenberg, II. Physiologisches Institut, Universität des Saarlandes, D-6650 Homburg (F.R.G.)

Temperature dependence of verapamil action

H. M. Piper[1]), R. Koch[2]) and P. G. Spieckermann[1])

[1]) Zentrum Physiologie, Abteilung Herzstoffwechsel, Universität Göttingen, Göttingen (F.R.G.)
[2]) Fachgebiet Medizinische Physik, Tierärztliche Hochschule Hannover, Hannover (F.R.G.)

Summary

Ca^{2+}-tolerant ventricular myocytes from adult rats were electrically stimulated. The maximal contraction frequency (f_m) was determined at different temperatures. In drug-free Tyrode solution, f_m follows the Arrhenius equation from 7 to 39.5 °C. However, verapamil introduces a discontinuity around 27 °C into the Arrhenius plot of f_m. Above this transition temperature the calcium antagonist lowers f_m more pronouncedly than below. Below, a tenfold higher concentration is needed for the same relative effect as at 37 °C. It is argued that this finding might be important in cardiac surgery when calcium antagonists are used for cardioplegia at deep hypothermia.

Key words: temperature dependence, verapamil, cardioplegia, hypothermia, adult cardiac myocytes

Introduction

Calcium antagonists are known to reduce contraction force of the heart without abolishing the generation of action potentials even at extremely high doses. Probably primarily due to the saving of contraction energy, cardiac energy requirements are lowered in the presence of calcium antagonists. This may be advantageous in ischemia (2, 3). Recently, the use of calcium antagonists, e.g. verapamil (12, 16), has been proposed as a new cardioplegic principle. At high doses these drugs can be applied for induction of reversible cardiac arrest. Ionic cardioplegic solutions are commonly applied in an advantageous combination with hypothermia. Therefore it seems necessary to investigate whether calcium antagonists are able to suppress elicited contractions, e.g. caused by surgical manipulations, equally well at low temperatures as at normothermia.

For quantifying the suppressing effects of calcium antagonists on the generation of contractions, we determined the maximal possible contraction frequency (f_m) under electrical stimulation at different temperatures. In isolated myocardial cells, f_m can be determined from the cell wall-movements even with a nearly complete force reduction and therefore may be regarded as a measure for the effective contractile refractoriness (17). The use of this pure myocyte system allows the exclusive attribution of the results to properties of the myocardial cell. These cells have high resting potentials and, when electrically stimulated, contract after elicitation of a normal action potential (11).

Methods

Ca^{2+}-tolerant ventricular myocytes from adult rats were prepared as previously described (13). The cells were incubated in aerobic Tyrode solution (NaCl 125.0 mM, KCl 2.6 mM, KH$_2$PO$_4$ 1.2 mM, MgSO$_4$ 1.2 mM, CaCl$_2$ 1.0 mM, glucose 5.0 mM, HEPES 10.0 mM at pH 7.4) with different concentrations of verapamil × HCl (Knoll, Ludwigshafen, F.R.G.). Response to electrical stimulation was followed by examining

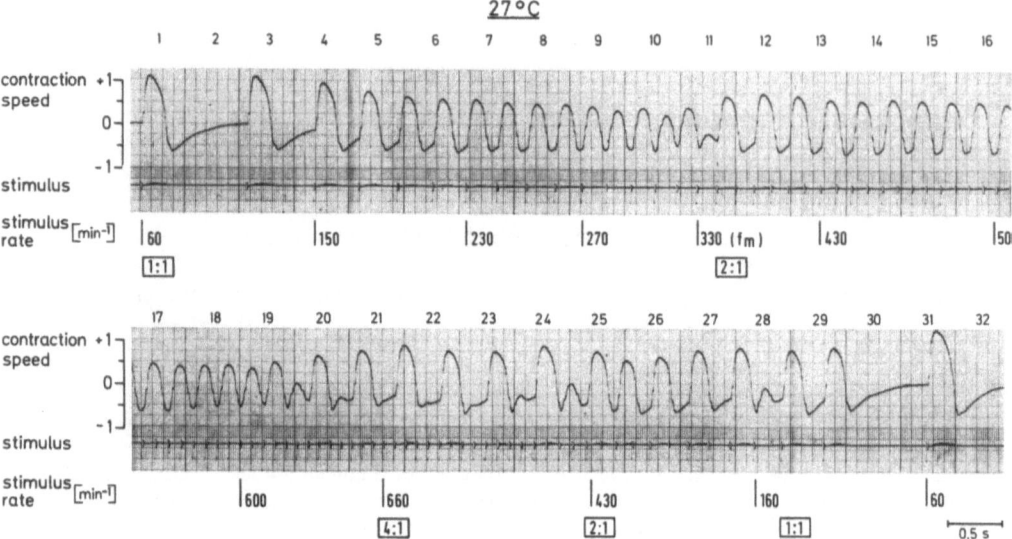

Fig. 1. Original registration of cell contractions under increasing and decreasing stimulation rate (contraction speed in arbitrary units) in drug-free Tyrode solution. The maximal rate, followed 1:1, is around 330 min⁻¹. Every second stimulus is followed above 330 min⁻¹, every fourth above 660 min⁻¹. The behaviour under decreasing stimulus rates exhibits hysteresis.

VERAPAMIL

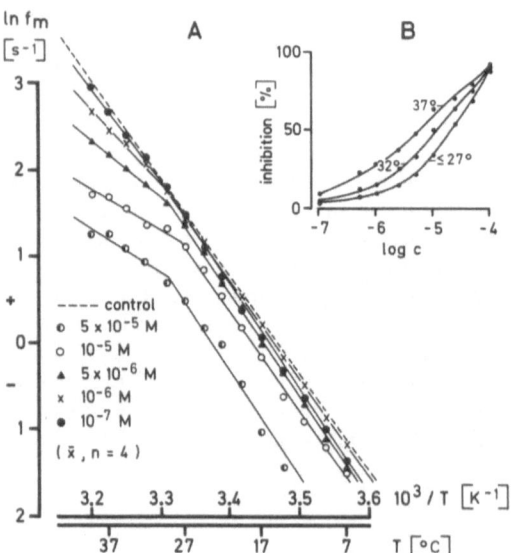

Fig. 2 A. Arrhenius plots of the maximal possible contraction frequency (f_m) under control conditions (broken line) and in the presence of verapamil (solid lines) (mean values of five independent experiments).
B. Dose-response curves for the effect of verapamil on f_m at different temperatures (in °C), expressed as percent inhibition relative to f_m under control conditions at the given temperature.

the cells under a phase contrast microscope with a temperature controlled stage. The frequency of cell contractions was determined with an opto-electronical detector (6) monitoring the light intensity modulations caused by the movements of a cell, part of which is seen in the visual field. In each experiment 20 cells of a given preparation were investigated. Rectangular voltage steps of 0.5 ms duration in trains of 10 pulses were applied to the electrolyte solution with platinum wire electrodes positioned 20 mm apart. Stimuli were adjusted to twice threshold intensity at medium temperature (22 °C). Then they were of over-threshold intensity at all temperatures. In order to minimize the influence of use-dependence and of diastolic tension development, repetition of pulse trains was only allowed after 5–10 min breaks sufficient for full recovery of the initial response. By this method the maximal contraction frequency (f_m) at a given temperature was determined. f_m was defined as the highest stimulation frequency followed 1:1 (Fig. 1). In each experiment a given cell sample was first cooled to 7 °C and then stimulated at stepwise raised temperatures. Raising the temperature was performed slowly (10 min per 2.5 °C). As proven in preliminary experiments the direction on the temperature scale has no influence on the results when the temperature is not changed too fast.

Results and discussion

The maximal contraction frequency f_m is a function of temperature with a Q_{10} value around 3.5 between 7 °C and 39.5 °C. As has been reported previously (13), f_m follows the Arrhenius equation ($\ln f_m = A - E/RT$, with constants A, E, R and absolute temperature T) in the investigated range from 7 °C to 39.5 °C. At 37 °C the stimulation rate can be increased to about 1000 min^{-1}, i.e., 2–3 times the heart rate of resting rats.

As demonstrated by Fig. 2A, verapamil lowers f_m in high concentrations at all temperatures. However, the effect is greater above a transition temperature around 27 °C. While below that temperature the slopes of the Arrhenius plots are the same as that of control, the slopes decline at higher temperatures under the influence of the calcium antagonist. At 10^{-5} M verapamil the Q_{10} value between 37 °C and 27 °C has declined from 3.5 (control) to 2. Thus, dose-response curves in a relative representation (Fig. 2B) differ according to temperature. In the range between 10^{-7} M and 10^{-5} M a given drug concentration at 37 °C has the same relative effect as the tenfold at temperatures below 27 °C. The procedure of short-time stimulation renders an influence of use-dependence on the observed behaviour unlikely. This is because even calcium antagonists with marked use-dependence for slow Ca^{2+} channel inhibition do not uncouple excitation-contraction in the strict sense. Contractile excitability does not disappear before almost total suppression of electrical excitability (7). f_m as a measure for contractile excitability does not depend on the contraction depth as long as this is not zero. In normothermia, concentrations effective on f_m are similar in magnitude to those diminishing the slow inward current and to those with a negative inotropic effect (1), a coincidence which does not hold for other calcium antagonists (1, 15). At very high concentrations, i.e. at 10^{-5} M to 10^{-4} M, a calcium antagonist may not only interfere with the sarcolemmal slow Ca^{2+} channel, but may have additional effects on other ion channels and on the intracellular mechanisms of electromechanical coupling. However, direct effects of verapamil on the calcium accumulation activity of mitochondria and on the tension development of the contractile proteins are improbable even at this dosage (4, 10).

For the time being, a molecular interpretation of the temperature dependence can only be postulated. Discontinuities in Arrhenius plots can often be related to physical transitions in the lipid phase of membranes. Under control conditions, f_m may follow the Arrhenius equation simply because the rate limiting reaction does. But in the presence of the calcium antagonist, a membrane bound enzyme may determine the overall rate, whose activity is dependent on the lipid environment under these circumstances. The relative lipophilic character of verapamil might speak in favour of this suggestion. Recently, evidence has been given for shielding of a polar head group of a sarcolemmal protein by the verapamil molecule (8). From this it might be suggested that the protein complex, after binding with verapamil, changes its interaction with its lipid environment.

If the hypothesis concerning the rate limiting steps holds, then a further statement can be made: below the transition temperature, the calcium antagonist has no influence on the activation energy of the rate limiting reaction. This is because the slope of the Arrhenius plot represents the activation energy of the reaction (ln k = A − E/RT, with reaction rate k and activation energy E). There are several indications that around 25 °C the sarcolemma of non-hibernating mammals undergoes a phase transition (5, 9). This might be the cause for spontaneous ventricular fibrillations occurring in bigger mammalian hearts below 28 °C (14). However, a shift in the Arrhenius plot would also be compatible with a protein conformation change caused by the drug at the transition temperature.

The fact that verapamil does not effectively suppress stimulatability of contractions in hypothermia renders its aptitude as a cardioplegic agent questionable, since the effect of other cardioplegic methods can even be potentiated by lowering the temperature. However, even if the specific cardioplegic effect is diminished, the calcium antagonist may be useful for the resuscitation of hearts from deep-hypothermic ischemia. First, their vascular effects may improve reperfusion conditions (2). Second, a pre-ischemic administration, even of low doses, may confer protection additional to hypothermia by lowering pre-ischemic energy turnover.

Acknowledgements

This study was supported by the Deutsche Forschungsgemeinschaft, SFB 89 – Kardiologie Göttingen. We wish to thank B. Eickhoff, H. Haacke and E. Neumeyer for their skilful technical assistance.

References

1. Bayer R, Kaufmann R, Rodenkirchen R (1982): The action of specific Ca antagonists on cardiac electrical activity. Progr Pharmacol 5: 53–85
2. Bourdillon PD, Poole-Wilson PA (1982): The effects of verapamil, quiescence and cardioplegia on calcium exchange and mechanical function in ischemic rabbit myocardium. Circ Res 50: 360–368
3. Fleckenstein A (1971): Specific inhibitors and promotors of calcium action in the excitation – contraction coupling of heart muscle and their role in the prevention or production of myocardial lesions. In: Harris P, Opie L (eds) Calcium and the Heart, pp 135–188, Academic Press, New York
4. Frey M, Janke J (1979): The effect of organic Ca-antagonists (verapamil, prenylamine) on the calcium transport system in isolated mitochondria of rat cardiac muscle. Pflügers Arch, Suppl to Vol 359: R 26
5. Gordon LM, Sauerheber RD, Esgate JA (1979): Spin label studies on rat liver and heart plasma membrane: effects of temperature, calcium and lanthanum on membrane fluidity. J Supramol Struct 9: 299–326
6. Koch R, Piper HM (1984): An opto-electronical system for analyzing contraction patterns of single myocardial cells. Pflügers Arch, Suppl to Vol 400: R 56
7. Koidl B, Tritthart HA (1982): D-600 blocks spontaneous discharge, excitability and contraction of cultured chick heart cells. J Mol Cell Cardiol 14: 251–257
8. Mannhold R, Steiner R, Rodenkirchen R, Bayer R (1983): Polarizability and dipole moment: determination and molecular importance for the verapamil action. Naunyn-Schmiedeberg's Arch pharmacol, Suppl to Vol 322: R 4
9. McMurchie EJ, Raison JK, Cairncross KD (1973): Temperature-induced phase changes in membranes of heart: A contrast between the thermal response of poikilotherms and homeotherms. Comp Biochem Physiol 44 B: 1017–1026
10. Nayler WG, Poole-Wilson PA (1981): Calcium antagonists: definition and mode of action. Basic Res Cardiol 76: 1–15
11. Pelzer D, Trube G, Piper HM (1984): Low resting potentials in single isolated heart cells due to membrane damage by the recording microelectrode. Pflügers Arch, 400: 197–199
12. Pinsky MW, Lewis RM, McMillin-Wood JB, Hara H, Hartley CJ, Gilette PC, Entman ML: Myocardial protection from ischemic arrest: potassium and verapamil cardioplegia. Am J Physiol 240: H 326–H 335

13. Piper HM, Probst I, Schwartz P, Hütter JF, Spieckermann PG (1982): Culturing of calcium stable adult cardiac myocytes. J Mol Cell Cardiol 14: 397–412

14. Popovic V, Popovic P (1974): Hypothermia in Biology and in Medicine. Grune and Stratton, New York

15. Raschak M (1976): Differences in the cardiac actions of the calcium antagonists verapamil and nifedipine. Arzneim-Forsch 26: 1330–1333

16. Robb-Nicholson C, Currie WD, Wechsler AS (1978): Effects of verapamil on myocardial tolerance to ischemic arrest. Circulation 58, Suppl 1: 119–124

17. Vaughan Williams EM, Szekeres L (1961): A comparison of tests for antifibrillatory action. Brit J Pharmacol 17: 424–432

Author's address:

H. M. Piper, Zentrum Physiologie, Universität Göttingen, Humboldtallee 23, D-3400 Göttingen (F.R.G.)

Microtubules and desmin filaments during the onset of heart growth in the rat

L. Rappaport, J. L. Samuel, B. Bertier-Savalle, F. Marotte and K. Schwartz

Unité 127 INSERM, Hôpital Lariboisière, 41 Bd de la Chapelle, Paris (France)

Summary

Microtubules and intermediate desmin filaments were visualized in rat cardiomyocytes during the onset of heart growth by double immunolabelling of isolated myocytes, with specific antibodies raised against tubulin and desmin. Heart growth was stimulated either by mechanical overloading induced by aortic stenosis, or by injection of thyroxine into hypothyroid rats. In both experimental models, alterations in the microtubule pattern were observed soon after stimulation of growth whereas desmin filament organization remained unchanged. Microtubules were redistributed in arrays parallel to the long axis of the myocytes and were more numerous around the nuclei. Microtubules therefore appeared to be involved in the cellular events occurring during stimulated heart growth irrespective of the nature of the stimulus.

Key words: microtubules, heart myocytes, immunofluorescence, growth

Introduction

Microtubules and intermediate desmin filaments are believed to be involved in muscle growth (1, 3, 7) and to contribute to the maintenance of cellular and myofibrillar organization during the contraction and relaxation cycles of striated muscle (4, 6, 7). On the basis of these properties we investigated the possible functions of microtubules and intermediate desmin filaments when heart muscle growth was stimulated by two different factors: one was mechanical (pressure overload induced by aortic stenosis (12)) and the other hormonal (treatment of hypothyroid rats with thyroxine). Both forms of stimulation enhance muscle work.

Methods

Details of the models and methods utilized have been published elsewhere (8, 9). Briefly, growth of heart muscle was stimulated either by pressure overload induced by stenosis of the aorta in 3-week old rats (2), or 4 µg/day of Thyroxine (T_4) was injected into 10-day old hypothyroid rats. Hypothyroidism was induced by an iodine-free diet containing propylthiouracil (PTU) during pregnancy and nursing. Myocytes were isolated by a perfusion method, purified on Ficoll, permeabilized in 1% Triton-X-100 in a 10 mM EGTA buffer, and fixed with 3.7% formaldehyde (8, 9). Microtubules and intermediate filaments were stained by double indirect immunolabelling with specific antibodies raised against brain tubulin (9) or gizzard desmin (5, 10). Tests of specificity included immunoblots, labelling of various types of cells, and the ELISA test.

Results

Heart growth was rapidly stimulated in both experimental models. (Fig. 1).

In normal rats, microtubules were seen to be organized around the nuclei, with concentrations at the poles and cones, and extensions in the cytoplasm in the form of loosely organized loops (Fig. 2b). This network was highly sensitive to depolymerizing agents such as colchicine and cold

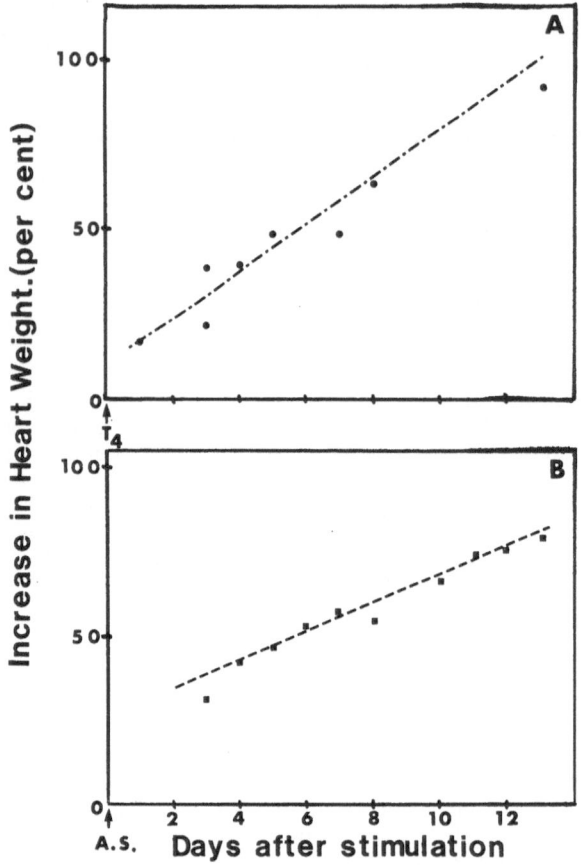

Fig. 1. Stimulation of heart growth in rat. Heart muscle growth was induced either by aortic stenosis in 3 week-old rats (B) or by injection of 10 day-old hypothyroid rat with 4 µg/day of T4 (A). A 50 per cent increase in growth was observed in both models about 7 days after stimulation.

(8). After 3 days, *aortic stenosis in 3 week-old rats* induced a densification of the microtubules, which were organized in arrays along the axis of the myocytes (Fig. 2a). The size of the myocytes did not change significantly. However, the increase in the weight of the heart (Fig. 1) observed concomitantly with a shift in the isomyosins pattern within myocytes (10), afforded proof of the adaptational processes occurring soon after aortic stenosis.

In thyroxine – deficient rats, myocyte growth was clearly delayed (Fig. 2e, f,). Myocytes were small and most of them displayed only one nucleus. Microtubules, concentrated around the nucleus, were underdeveloped throughout the cytoplasm, where they appeared as straight rod-shaped filaments (Fig. 2e).

After 3 to 5 days *injections of thyroxine* induced an increase in the size of the myocytes as well as a reorganization of the microtubule pattern similar to that described after aortic stenosis. (Fig. 2c).

Antidesmin stained the Z bands, but the intercalated disks only stained faintly as they had been partly torn out during myocyte isolation by collagenase (Fig. 2d, f). No change in either the distribution or intensity of the labelling was detected after stimulation of growth, whether by stenosis (10) or T4-treatment (Fig. 2d).

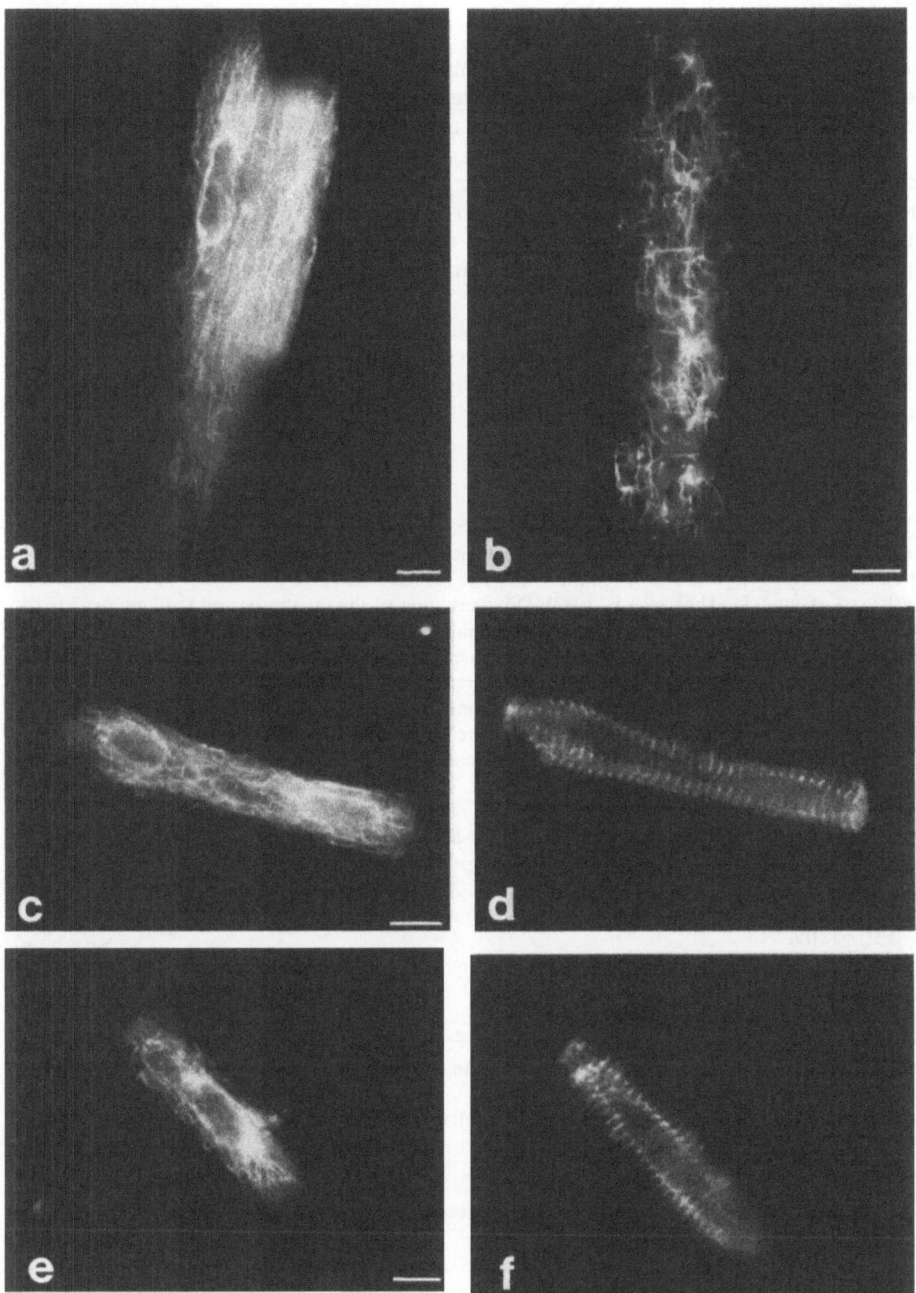

Fig. 2. Immunolabelling of cardiac myocytes with antitubulin and antidesmin during stimulated heart growth. Heart muscle growth was induced as in Fig. 1. In model (B) myocytes were isolated 10 days after aortic stenosis (a) or sham-surgery (b) and labelled with antitubulin. In model (A) myocytes were purified from 3 week-old hypothyroid rats treated with T4 for 9 days (c, d), or not treated (e, f). Myocytes were double labelled with antibulin (c, e) and antidesmin (d, f).

Discussion

Studies of microtubules in rat heart have been hampered by the fact that they are minor components of the myocytes (6, 9). Moreover, the relative thickness of heart myocytes (30 μm), as well as the diameter of the microtubules (25 nm) and their three-dimensional organization made it difficult to visualize the microtubule network on cryostat sections. In our hands the best approach was the one already described for cells in culture (11), i. e. immunostaining of tubulin in isolated myocytes permeabilized with Triton-X-100, in a buffer known to stabilize microtubule structures. Thus, by double immunolabelling, a minor myocyte component such as tubulin (10 μg/100 mg proteins) was visualized together with major components of sarcomeres like myosins (9) or of cytoskeleton, such as desmin.

Our results favour the idea that two different stimulations of heart muscle growth affect the microtubule pattern and not the organization of desmin. This reorganization of the microtubule pattern might be connected with the early events that occur during acute growth such as myofibrillogenesis, or multiplication of mitochondria.

References

1. Auber J (1969): La myofibrillogenèse du muscle strié. 1. Insectes. J de Microscopie 8: 197–232
2. Bugaisky LB, Siegel E, Whalen RG (1983): Myosin isozymes changes in the heart following constriction of the ascending aorta of a 25 day old rat. FEBS lett 161: 230–234
3. Carlsson E, Kjorell U, Thornell LE (1982): Differentiation of the myofibrils and the intermediate filament system during post-natal development of the rat heart. Eur J Cell Biol 27: 62–73
4. Ferrans VJ, Roberts SC (1973): Intermyofibrillar nuclear myofibrillar connections in human and canine myocardium: an ultrastructural study. J Mol Cell Cardiol 5: 247–258
5. Geisler N, Weber K (1980): Purification of smooth-muscle desmin and a protein-chemical comparison of desmins from chicken gizzard and hog stomach. Eur J Biochem 111: 425–433
6. Goldstein MA, Entman ML (1979): Microtubules in mammalian heart muscle. J Cell Biol 80: 183–195
7. Lazarides E (1978): The distribution of desmin (100 A) filaments in primary cultures of embryonic chick cardiac cells. Exp Cell Research 112: 265–273
8. Samuel JL, Rappaport L, Mercadier JJ, Lompre AM, Sartore S, Triban C, Schiaffino S, Schwartz K (1983): Distribution of myosin isozymes within single cardiac cells. Circ Res 52: 200–209
9. Samuel JL, Schwartz K, Lompre AM, Delcayre C, Marotte F, Swynghedauw B, Rappaport L (1983): Immunological quantitation and localization of tubulin in adult rat heart isolated myocytes. Eur J Cell Biol 31: 99–106
10. Samuel JL, Bertier L, Bugaisky L, Marotte F, Swynghedauw B, Schwartz K, Rappaport L (1984): Different distributions of microtubules, desmin filaments and isomyosins during the onset of cardiac hypertrophy in the rat. Eur J Cell Biol 34: 300–306
11. Schliwa M, Van Blerkon J, Porter K (1981): Stabilization of the cytoplasmic ground substance in detergent-opened cells and a structural and biochemical analysis of its composition. Proc Natl Acad Sci USA 78: 4329–4333
12. Swynghedauw B, Delcayre C (1982): Biology of cardiac overload. Pathobiol Ann 12: 137–138

Author's address:

Dr. L. Rappaport, Unité 127 INSERM, Hôpital Lariboisière, 41 Bd de la Chapelle, 75010 Paris (France)

The correlation between catecholamine and lipid peroxidation induced damage in heart cells

A. A. Noronha-Dutra, E. M. Steen, and N. Woolf

The Bland-Sutton Institute of Pathology, The Middlesex Hospital Medical School, London (U.K.)

Summary

Cardiocytes incubated with isoprenaline plasma showed an initial rise in the intracellular ATP levels followed by a steady fall; after two hours the ATP levels of the cells treated with the plasma were less than half that of the control cells. This fall in intracellular ATP was accompanied by morphological changes, in particular the formation of both large and small membrane blebs.

Key words: catecholamines, lipid peroxidation, necrosis, free radicals, sarcolemmal blebbing

1 Introduction

The cardiotoxicity of both isoprenaline and adrenaline is well known. At high concentrations these amines induce subendocardial necrosis (2), and isoprenaline in particular, produces changes which resemble the subendocardial laminar necrosis produced by myocardial ischemia in humans (3). As a result isoprenaline has been used extensively as a model of myocardial necrosis.

A number of theories have been presented to explain the pathogenetic mechanisms of isoprenaline necrosis, the majority of which tend to revolve around Ca^{2+} with cAMP occasionally appearing as supporting actor. The first hypothesis, ischemia, has been considered to occur as a result of the isoprenaline induced reduction in blood pressure due to peripheral vasodilation which in turn reduces the diastolic pressure. This reduction has been shown to decrease the coronary flow to the subendocardial area of the heart (1, 5).

The second hypothesis involving Ca^{2+} overload was suggested to result from the fact that high concentrations of catecholamine would induce a large rise in intracellular Ca^{2+} (an effect which can be mimicked by cAMP). In an attempt to buffer this extra Ca^{2+} the cells would deplete their high energy phosphate reserves until this buffering capacity was overcome, resulting in cell death (4).

Recently we have shown that isoprenaline and adrenaline at high concentrations (10^{-4} M) which are known to cause damage *in vivo* and in the presence of as much as 4 mM Ca^{2+} were not lethal to isolated heart cells and indeed no ultrastructural changes could be detected (6,9). Measurement of cell viability using carboxyfluorescein diacetate confirmed that no cell damage had occurred.

At the same time some parallel work *in vivo* gave results which showed that some factor other than ischemia was involved in the pathogenesis of isoprenaline necrosis. Examination by scanning and transmission electron microscopy of the endocardium and adjacent myocardium of the hearts of rats treated with isoprenaline and killed at different time intervals showed that the initial damage was to the endocardial cells (30 minutes) with the myocardium becoming involved later (8). Some of the morphological changes, in particular the crenated membranes of the myocardial cells, resembled changes seen in macrophage and fibroblasts after being irradiated or subjected to a severe free radical attack. Although it was doubtful whether isoprenaline could mimic such harsh conditions, these results suggested to us that the myodardium may be sensitive to free radical attack.

When isolated heart cells were treated with very low doses (0.1 mM) of cumene hydroperoxide, a compound that initiates lipid peroxidation through free radicals, it caused dramatic changes in these cells, in particular blebbing of the cell membrane and contraction; no change was seen in fibroblasts at this concentration. The same changes occurred when heart cells were treated with diamide (7) (a compound which preferentially oxidizes-SH groups of small molecules, mainly glutathione). Glutathione peroxidase acts as the cell's main defense aganst lipid peroxidation which would occur as a result of free radical attack.

The combined data prompted us to examine the effect of plasma taken from an isoprenaline treated rat on the isolated heart cells and to determine whether free radicals were present in the plasma. The following are some preliminary results.

Materials and methods

Blood was taken from male rats (Charles River U.K. strain Crl: CD(SD)BR) treated with isoprenaline bitartrate (10 mg/kg), rapidly separated by microcentrifugation, and the plasma added to the heart cells, prepared as previously described (7) (the plasma was added to the cells within three minutes of taking the blood). Control cells were incubated with plasma from an untreated rat and changes in cell ATP levels were determined using the firefly enzyme luciferin luciferase and an LKB Wallac luminometer 1250 as described previously (7). The presence of free radicals was detected by measuring chemiluminescence which is produced during lipid peroxidation by the energy decay of free radicals produced during this chain reaction, using equipment designed in our laboratory. It was necessary to do this as chemiluminescence measurement involves very low levels of light, mainly in the infra-red region, well beyond the capability of any commercially available instrument. (We immediately discarded the idea of using the thiobarbituric acid method which measures malondialdehyde, an end product of lipid peroxidation of some particular fatty acids, because this method is not sensitive enough and is full of pitfalls. The other method, diene conjugation, which we used before is also not sufficiently sensitive.)

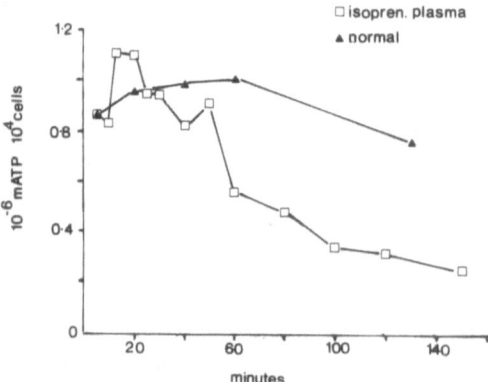

Fig. 1. ATP changes in cells treated with isoprenaline plasma and normal plasma.

Results

The cells incubated with isoprenaline plasma showed an initial rise in the intracelluar ATP levels followed by a steady fall; after two hours the ATP levels of the cells treated with the plasma were less than half that of the control cells (Fig. 1). This fall in intracellular ATP was accompanied by morphological changes, in particular the formation of both large and small membrane blebs (Fig. 2).

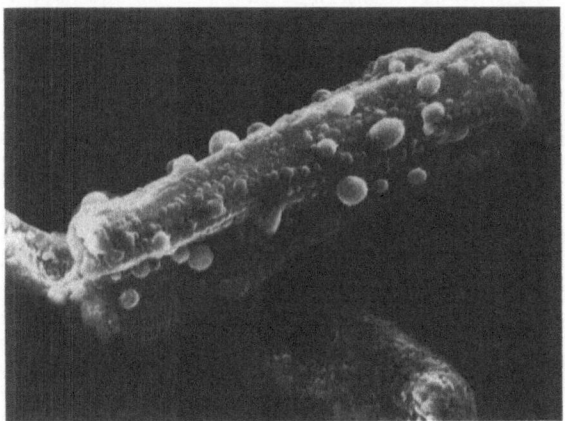

Fig. 2. Scanning electron micrograph of a heart cell which has been incubated with isoprenaline plasma for one hour. The cell surface is covered with both large and micro blebs. x1,000.

Our preliminary results of chemiluminescence measurements have shown the presence of free radicals in the plasma from isoprenaline treated rats (this blood was taken and treated as described previously); there were more than three times as many counts from the isoprenaline plasma compared to the normal plasma (Fig. 3).

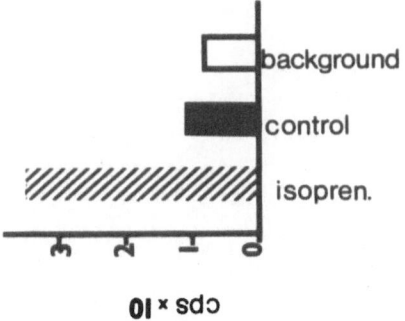

Fig. 3. Chemiluminescence of isoprenaline plasma.

From these initial chemiluminescence measurements and the similarities between heart cells treated with cumene hydroperoxide and diamide and those treated with isoprenaline plasma, in addition to our previous experiences with isoprenaline both *in vivo* and *in vitro*, it seems clear to us that free radicals and lipid peroxidation are involved in the process of isoprenaline induced damage of the heart.

References

1. Buckberg GD, Ross G (1973): Effects of isoproterenol on coronary blood flow: Its distribution and myocardial performance. Cardiovasc Res 7: 429–437
2. Chappel CI, Rona G, Balazs T, Gaudry R (1959): Severe myocardial necrosis produced by isoproterenol in the rat. Archs Int Pharmacodyn Ther 122: 123–128
3. Davies MJ (1977): The pathology of myocardial ischemia. J Clin Pathol 30 (Suppl II): 45–52
4. Fleckenstein A, Janke J, Doring JH, Leder O (1974): Myocardial fiber necrosis due to intracellular Ca-overload – a new principle in cardiac physiology. Recent Advances in Studies on Cardiac Structure and Metabolism, 4, Myocardial Biology, pp 536–580 University Park Press, Baltimore
5. Moir TW, DeBra D (1967): Effect of left ventricular hypertension, ischemia and vasoactive drugs on the myocardial distribution of coronary flow. Circ Res 21: 65–74
6. Noronha-Dutra AA, Steen EM (1981): The effects of isoprenaline and propranalol on ATP levels of isolated cardiac myocytes. (Abstr) J Pathol 134: 311
7. Noronha-Dutra AA, Steen EM (1982): Lipid peroxidation as a mechanism of injury in cardiac myocytes. Lab Invest 40: 183–193
8. Noronha-Dutra AA, Steen EM, Woolf N (1984): The early changes induced by isoproterenol in the endocardium and adjacent myocardium. Am J Path 114: 231–239
9. Steen E, Noronha-Dutra AA, Woolf N (1982): The response of isolated rat heart cells to cardiotoxic concentrations of isoprenaline. J Path 137: 167–176

Author's address:

A.A. Noronha-Dutra, Bland-Sutton Institute, Middlesex Hospital Medical School, Riding House St., London W1 7PN (U.K.)

Anoxia in neonatal rat heart cell cultures[1])

A. Van der Laarse, J. C. Altona and C. H. F. Bloys van Treslong

Laboratory of Cardiobiochemistry, University Hospital, Leiden (The Netherlands)

Summary

Monolayer cultures of heart cells are prepared by dissociation of neonatal rat hearts with collagenase. The regularly and synchronously contracting monolayer is subjected to oxygen and metabolic substrate deprivation for some time (anoxia), and is, in a number of experiments, followed by a short period of oxygen and metabolic substrate repletion (reoxygenation). Analysed were the frequency and regularity of beating, number of nonvital cells, and enzyme activities and DNA content in the cells as well as in the extracellular medium. We observed that a correlation exists between the released activity of a cytoplasmic enzyme, α-hydroxybutyrate dehydrogenase (HBDH) and i) number of nonvital cells, ii) depression of beating frequency measured during reoxygenation, iii) the released activities of enzymes from sarcolemma (L-leucylnaphthylamidase), from lysosomes (N-acetyl-β-glucosaminidase), and mitochondrial outer membrane (monoamine oxidase). No correlation exists between the released activity of HBDH and a) the released activity of an enzyme system from the mitochondrial inner membrane (succinate: cytochrome c reductase), and b) the released amount of DNA. Futhermore, reoxygenation of anoxic heart cell cultures leads to a suddenly occurring HBDH release which phenomenon is known as "oxygen paradox".

Key words: enzyme release, cell beating, trypan blue, anoxia, oxygen paradox

Introduction

Cardiovascular research on monolayer cultures of neonatal rat heart cells was started in 1960 by Harary and coworkers (3). They studied the effects of cardiac drugs, metabolic substrates and metabolic inhibitors on beating rate of the cultures. DeLuca et al. (1) were the first to study the effects of oxygen deprivation on cell survival in this preparation. In 1976, our research group started a similar study to investigate at a cellular level the mechanisms and characteristics of cardiac cell death. Particularly, we were interested in the intracellular disturbances brought about by metabolic energy deprivation.

Methods

Monolayer cultures of heart cells from 2-day old rats were prepared according to Van der Laarse et al. (9). Cultures grown for 3 days were used for investigation. To that purpose, the culture medium was replaced by a modified Krebs-Henseleit bicarbonate buffer (KHB), containing no metabolic substrates, or the substrate(s) indicated. The cultures were investigated in a thermostated chamber which is mounted on the stage of an inverted microscope (Leitz Diavert) with phase-contrast optics (10). Cellular pulsation was recorded opto-electrically (EMI photomultiplier). After a 1 h period of pre-incubation of the culture in KHB containing pyruvate (10 mmol/l) gassed with 95% air + 5% CO_2, the medium was replaced by KHB without substrates. This buffer was pre-gassed with 95% N_2 + 5% CO_2. During anoxic incubation the incubation chamber was

---

[1]) The investigations were supported in part by the Foundation for Medical Research FUNGO, which is subsidised by the Netherlands Organisation for the Advancement of Pure Research (Z.W.O.).

flushed with 95% N_2 + 5% CO_2. Reoxygenation was introduced by replacing the buffer by KHB containing pyruvate (10 mmol/l), pre-gassed with 95% air + 5% CO_2. Gas flush through the chamber was switched to the same gas mixture. During incubation, samples of the medium can be drawn for biochemical investigation. At the end of incubation the monolayer is *either* stained for nonvital cells by trypan blue (2%), *or* taken up in an aliquot of phosphate-buffered saline, scraped from the dish, homogenised in a Potter tube and sonified (Branson Sonic Power Corp.). Medium samples and the cellular homogenate are assayed for several enzyme activities and for concentration of DNA (4). Assayed enzymes are α-hydroxybutyrate dehydrogenase (HBDH) (7), N-acetyl-β-glucosaminidase (NAGA) (6), L-leucylnaphthylamidase (LNA) (2), monoamine oxidase (MAO) (5) and succinate: cytochrome c reductase (SCR) (8). HBDH is used as a reference enzyme. To that purpose the HBDH activity of the culture at t = 0 is calculated according to the following formula:

cell. HBDH act. (t = 0) = cell. HBDH act (t) + released HBDH act. (t) (eq. 1).

Instead of an expression per culture or per mg protein, we prefer to express cellular activities and concentrations per unit (U) HBDH at t = 0. Therefore, cellular activities and concentrations at time t are divided by the cellular HBDH activity in the same culture at time t. Via eq. 1 the ratio of cellular activity (t) or concentration (t) per cellular HBDH activity (t = 0) is obtained.

Results

Transient anoxia and contractile function

After transient anoxia a fall in frequency of beating and an increased irregularity of beating are observed, both effects being dependent on the duration of anoxia (Fig. 1). These findings indicate the occurrence of an anoxia-induced decrease of electrical coupling between myocytes, either by an effect of the gap junction itself, or by cell death of a number of myocytes leading to decreased coupling between surviving myocytes.

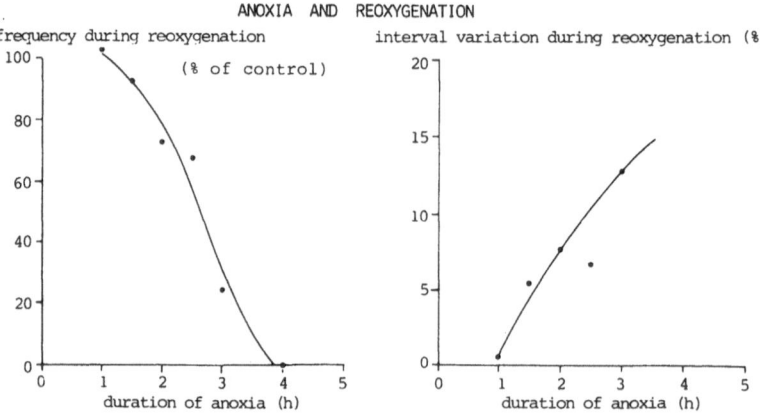

Fig. 1. Effect of a period of transient anoxia on frequency of beating during reoxygenation and on regularity of beating during reoxygenation, expressed as the variation coeffficient of 10 successive intervals between beats.

Transient anoxia, cell death and enzyme release

Anoxia causes liberation of enzymes from myocytes into the surrounding fluid. The released activity of a cytoplasmic enzyme, HBDH, expressed as % of the total activity of HBDH in the culture at t = 0, is equal to the % of necrotic cells in the culture as measured with the use of trypan blue, at any time t during anoxic incubation. The % fall of spontaneous beating frequency

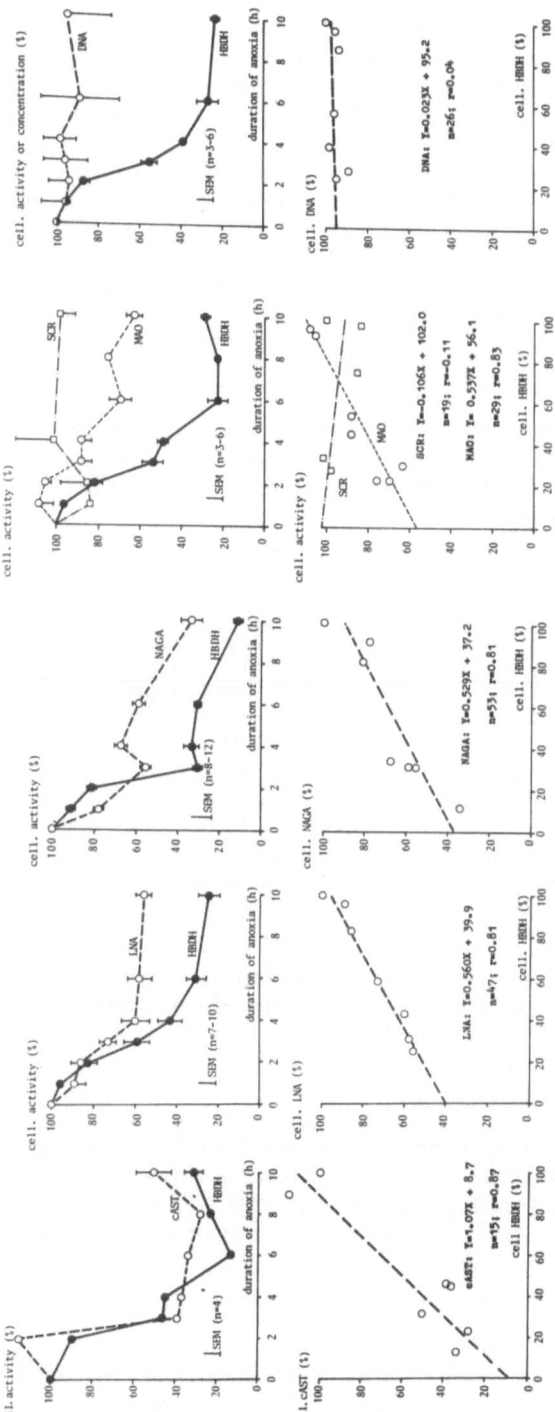

Fig. 2. Top: Time-course of intracellular activity or intracellular concentration of, from left to right, CAST, LNA, NAGA, SCR and MAO, and DNA during 10 h of anoxic incubation of heart cell cultures. In each panel intracellular HBDH activity is presented, which enzyme is used as reference. Bottom: Relation between intracellular activity or intracellular concentration of, from left to right, cAST, LNA, NAGA, SCR and MAO and DNA and intracellular HBDH activity. Circles indicate mean values of data obtained at equal durations of anoxia. Number of data is given by n. Linear regression lines are drawn. Abbreviations are explained in the text.

after a period of transient anoxia was proportional to the duration of anoxia and the % HBDH released. When after varying periods of anoxia, cellular HBDH activities are compared with cellular activities of other enzymes (or concentrations of other macromolecules) an impression is obtained of the extent to which the enzymes (or macromolecules) are liberated from the cultures. Ultimate HBDH release is approximately 100%. After linear regression of a cellular activity (or concentration) to the cellular HBDH activity at different times t during anoxic incubation, extrapolation of the regression line to a cellular HBDH activity of zero gives an estimate of the unreleased activity (or concentration) of that particular enzyme (or macromolecule). Fig. 2 shows time-courses and regression relations for cAST,LNA, NAGA, MAO, SCR and DNA, with HBDH as the reference enzyme. The unreleasable portion of DNA and SCR is roughly 100%, for MAO, NAGA and LNA 40–60%, and for cAST 0%.

Factors modifying anoxia-induced enzyme release.

Factors which partially inhibit HBDH release from anoxic cultures are the presence of glucose and, independently, a low (0.025 mmol/l) Ca^{2+}-ion concentration (Table 1). Contrarily, reoxygenation leads to an exacerbation of HBDH release, although reoxygenation after 1, 2 and 3 h of anoxia will save a number of cells from dying.

Table 1. Anoxia-induced HBDH release.

Duration of anoxia (h)	Activity of HBDH released (% of cellular HBDH activity at t = 0)		
	2.5 mmol/l Ca^{2+} no glucose	0.025 mmol/l Ca^{2+} no glucose	2.5 mmol/l Ca^{2+} 5 mmol/l glucose
2	4.0 ± 0.9 (5)	3.0 ± 0.3 (4)	1.1 (1)
3	21.8 ± 2.7 (28)	4.4 ± 0.3 (4)*	2.6 (1)
4	51.6 ± 2.8 (28)	6.2 ± 0.7 (4)**	2.7 (1)
5	69.9 ± 1.9 (27)	8.5 ± 0.3 (4)**	5.6 (1)

Indicated are mean values $\pm$ SEM; n is number of observations, given within parentheses. *$p < 0.05$ and ** $p < 0.01$, compared to anoxia in the presence of 2.5 mmol/l Ca^{2+} and in the absence of glucose.

Reoxygenation and enzyme release

Reoxygenation of anoxically incubated cultures induces an immediate release of enzymes from the cultures. Fig. 3 shows the activity of HBDH released in the first h of reoxygenation after 1–6 h of anoxia, and, for comparison, in 1 h of continued anoxia.

Factors modifying reoxygenation-induced enzyme release

Exacerbation of enzyme release caused by reoxygenation is completely absent in the presence of KCN (5 mmol/l) during reoxygenation (Table 2). Then, no difference in enzyme release is observed between prolonged anoxia and reoxygenation in the presence of KCN. Reoxygenation in the presence of a low (0.025 mmol/l) Ca^{2+}-ion concentration reduces enzyme release during reoxygenation (Table 2).

Discussion

The studies discussed and our results presented show that the model of anoxic rat heart cell cultures can be used to investigate a wide number of cellular changes, often seen in whole hearts under several pathological conditions. As the cultures are easy to handle and have a simple con-

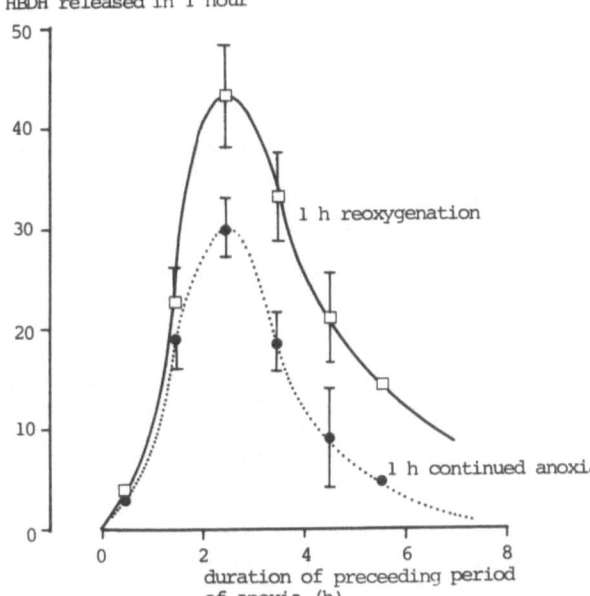

% HBDH released in 1 hour

Fig. 3. Relation between released activity of α-hydroxybutyrate dehydrogenase (HBDH), expressed as % of intracellular HBDH activity at t = 0, in 1 h of reoxygenation (□—□) or in 1 h of continued anoxia (● ... ●) and duration of preceding period of anoxia.

figuration (monolayer) independent of any arterial supply and venous drainage, the culture model offers many advantages over whole hearts in elucidating specific pathophysiological changes in cardiac cells.

Table 2. Reoxygenation-induced HBDH release.

Duration of anoxia (h)	Activity of HBDH released (% of cellular HBDH activity at t = 0)			
	1 h of continued anoxia	1 h of reoxygenation	1 h of reoxygenation in the presence of KCN (5 mmol/l)	1 h of reoxygenation
		Ca^{2+} = 2.5 mmol/l	Ca^{2+} = 2.5 mmol/l	Ca^{2+} = 0.025 mmol/l
2	19.1 ± 3.1 (23)	23.1 ± 2.9 (5)	17.1 ± 6.0 (5)	16.6 ± 0.8 (3)
3	29.8 ± 2.5 (28)*	42.8 ± 4.5 (6)	28.9 ± 6.7 (5)*	27.8 ± 3.6 (5)*
4	18.3 ± 3.1 (27)*	33.0 ± 4.2 (6)	19.9 ± 5.8 (5)*	17.3 ± 3.4 (5)**
5	9.0 ± 4.9 (5)*	21.1 ± 4.3 (6)	10.4 ± 3.0 (5)*	12.3 ± 2.0 (4)*
6	4.7 (1)	14.6 (1)		

Indicated are mean values ± SEM; n is number of observations, given within parentheses. * p < 0.05 and ** p < 0.01, compared to 1 h of reoxygenation in the presence of 2.5 mmol/l Ca^{2+}.

References

1. DeLuca MA, Ingwall JS, Bittl JA (1974): Biochemical response of myocardial cells in culture to oxygen and glucose deprivation. Biochem Biophys Res Commun 59: 749–756
2. Goldbarg JA, Ruthenburg AM (1958): The colorimetric determination of leucine aminopeptidase in urine and serum of normal subjects and patients with cancer or other diseases. Cancer 11: 283–293
3. Harary I, Farley B (1960): In vitro studies of single isolated beating heart cells. Science 131: 1674–1675
4. Karsten U, Wollenberger A (1977): Improvements in the ethidiumbromide method for direct fluorimetric estimation of DNA and RNA in cell and tissue homogenates. Anal Biochem 77: 464–470
5. Krajl M (1965): A rapid microfluorimetric determination of monoamine oxidase. Biochem Pharmacol 14: 1684–1685
6. Leaback DH, Walker PG (1961): Studies on glucosaminidase. 4. The fluorimetric assay of N-acetyl-β-glucosaminidase. Biochem J 78: 151–156
7. Rosalki SB, Wilkinson JH (1960): Reduction of α-ketobutyrate by human serum. Nature 188: 1110–1111
8. Stumpf DA, Parks JK (1981): Human mitochondrial electron transport chain: Assay of succinate: cytochrome c reductase in leukocytes, platelets and cultured fibroblasts. Biochem Med 25: 234–238
9. Van der Laarse A, Hollaar L, Kokshoorn LJM, Witteveen SAGJ (1979): The activity of cardio-specific isoenzymes of creatine phosphokinase and lactate dehydrogenase in monolayer cultures of neonatal rat heart cells. J Mol Cell Cardiol 11: 501–510
10. Van der Laarse A, Hollaar L, Van der Valk LJM (1979): Release of alphahydroxybutyrate dehydrogenase from neonatal rat heart cell culture exposed to anoxia and reoxygenation: comparison with impairment of structure and function of damaged cardiac cells. Cardiovasc Res 13: 345–353

Author's address:

Dr. A. Van der Laarse, Lab. Cardiobiochemistry, Building 20, University Hospital Leiden, Rijnsburgerweg 10, 2333 AA Leiden (The Netherlands)

Enzyme release and glycolytic energy production

H. M. Piper, R. Spahr, J. F. Hütter, and P. G. Spieckermann

Zentrum Physiologie, Abteilung Herzstoffwechsel, Universität Göttingen, Göttingen (F.R.G.)

Summary

In substrate-free anoxia, activities of released cytosolic enzymes (LDH, MDH) correlate inversely with the actual ATP level (for both: r = −0.98). At the same time there is a close correlation between lactate production from glycogen and the ATP content (r = 0.98). With external glucose present enzyme release is greatly delayed, but this could be due to the stimulation of glycolysis as well as to the maintenance of high ATP levels. When glycolysis is blocked by iodoacetate under aerobic conditions, the cells also become depleted of high-energy phosphates. This depletion is delayed in the presence of pyruvate. Cytosolic enzyme release again is correlated with total ATP contents, by the same relation in the presence or absence of pyruvate. Glycolytic energy production is negligible in both cases and does not seem to determine enzyme release directly.

Key words: anoxia, glycolysis, adenosine triphosphate, enzyme release, adult cardiac myocytes

Introduction

It has been hypothesized by Bricknell and Opie (3) that glycolytically generated ATP plays a major role in the maintenance of cell membrane integrity under conditions of energy depletion. Indeed, they found that in low flow ischemia with several substrates those allowing higher glycolytic flux rates led to a lower enzyme release, even if compared at identical homogenate ATP-levels. It was also observed by the same group that contracture development at a given total ATP content is less when glycolytic flux is high (2). Therefore the question might be asked whether higher enzyme release at low rates of glycolysis is due to compartmentalized availability of ATP or secondary to contracture-induced cell damage. Since in tissue contracture may cause cell damage by mechanical cell-cell interaction, and such forces do not occur in an isolated cell system, use of isolated myocytes may help to differentiate between the aforementioned alternatives. It has been stated elsewhere that in isolated myocytes contracture development leads, not immediately, to cytolysis (10).

Methods

A cell culture of intact, Ca^{2+}-stable ventricular muscle cells from 12 weeks old Sprague-Dawley rats was prepared as previously described (9). 4 hours after plating, the dishes (60 mm, Falcon) were washed and filled with 1 ml of a modified anoxic Tyrode solution (125.0 mM NaCl, 2.6 mM KCl, 1.2 mM KH_2PO_4, 1.2 mM $MgSO_2$, 1.0 mM $CaCl_2$, 25.0 mM HEPES, pH 7.4, equilibrated with 100% N_2 at 37 °C) or with an aerobic solution of the same composition with additions of 10 mM glucose and 1 mM iodoacetate (IAA) ± 5 mM pyruvate. Dishes, containing about 10^5 cells, were then transferred to an incubation chamber that was gently perfused with water-saturated nitrogen or air and slowly rotated (1 cycle/5 min) at a slight inclination. Under nitrogen or air, supernatant was withdrawn and perchloric acid was added. Oxygen pressure was found less than 1 mm Hg after 5 min under N_2, pH was between 7.35 and 7.4 in all experiments. CP, ATP, glycogen (as glucose), lactate, fructose-1,6-diphosphate and activities of lactate dehydrogenase (LDH, EC 1.1.1.27), ma- late dehydrogenase (MDH, EC 1.1.1.37) and glutamate dehydrogenase (GLDH, EC 1.4.2.3) were determined by standard enzymatic UV-methods (1). Acid phosphatase (acid P'ase, EC 3.1.3.2) activity was assayed with 4-nitrophenylphosphate as substrate, as described in (1). Enzyme activities are expressed as a percentage of the initial cellular content determined in cells from control samples which were sonicated in 1 ml Tyrode with 1% Triton X-100. Incubated under air in the plain salt buffer or in the buffer plus glucose, cells did not change high-energy phosphate contents nor lose significant enzyme activities during 120 min. If not stated differently, each datum represents the mean ± standard deviation of 5 independent experiments.

Results and discussion

Under substrate-free anoxic conditions lactate production rate decreases early after the initial stimulation caused by the transition to anaerobiosis (initial rate: 1.2 ± 0.1 µmol lactate/g_{ww} × min). Although only $48 \pm 4\%$ of the initial glycogen contents are consumed after 60 min ano- xia, the glycolytic flux has then fallen to one third of its inital value. Since high-energy phosphate contents decrease concurrently, lactate production rate is found correlated to the actual ATP content (Fig. 1).

It is not yet fully known why, in this system as well as in the substrate-free anoxic perfused heart (5), glycolysis already begins to decelerate at high ATP levels. The correlation between lac- tate production and the ATP level may reflect a causal relation, or may be a secondary phenome- non, e.g. due to alterations of the ion balances in the anoxic cell (6). In the oxygen deficient heart,

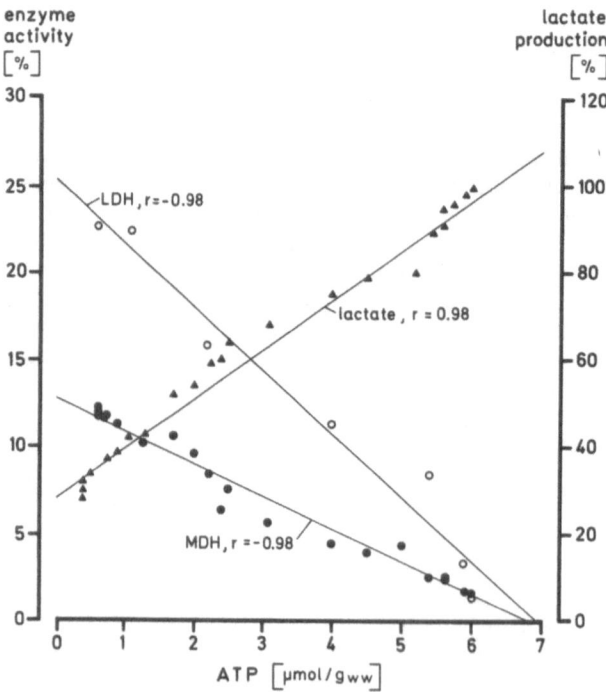

Fig. 1. Substrate-free anoxia: relation of released activities of LDH and MDH (expressed as a percentage of the initial total activity) and of lactate production rate (expressed as a percentage of the initial rate: 1.2 µmol/ g_{ww} × min) to the actual ATP content. Each datum represents the mean value of 5 independent experi- ments.

inhibition of glycolysis near an average ATP level of 2 $\mu mol/g_{ww}$ was attributed to progressive displacement of the mass action relationship of the phosphofructokinase reaction from equilibrium (8).

During the early phase of anoxia, characterized by a progressive decay of high-energy phosphates, a moderate release of cytosolic enzymes (LDH, MDH) was observed, while extracellular activities of mitochondria-specific GLDH or lysosomal acid phosphatase remained unchanged (10). Because the medium activities of cytosolic enzymes again correlate with the actual ATP levels (Fig. 1) there is also a correlation between released cytosolic enzyme activities and lactate production rate (for LDH and MDH $r = -0.97$, $p < 0.001$). But this means that the rate of enzyme release decreases similarly in time to the rate of lactate production, and it therefore does not support the hypothesis that low glycolytic activity goes along with high rates of enzyme release.

With exogenous glucose present, glycolysis is largely stimulated. When cells are incubated anaerobically in the rich M-199 medium, containing 5.5 mM glucose, lactate production is maintained constant at a high level for 120 min: 3.2 ± 0.2 $\mu mol/g_{ww} \times min$. Under these conditions, enzyme release is largely postponed. At 120 min anoxia only $1.7 \pm 0.2\%$ of total control activity of MDH is released. At this time, ATP-levels are still unaffected.

The experiments described so far did not allow to differentiate between the importance of glycolytic energy production and the presence of a certain total ATP-level, since states with low enzyme release exhibited both high rates of glycolysis and high ATP levels. Therefore we investigated enzyme release under conditions of energy depletion under blockade of glycolytic flow. IAA is an inhibitor of glycolysis. In principle it is a non-specific alkylating agent, however, at lower concentrations its inhibitory effect on glyceraldehyde-3-phosphate dehydrogenase is rather specific (4). Therefore, under aerobic conditions, ATP production can only be maintained when carbohydrates are available that may enter the Embden-Meyerhof pathway below this site of blockade, or when acetate or fatty acids are available. In the absence of exogenous substrates, IAA produces a rapid decline of high-energy phosphates, most probably because endogenous fatty acids cannot be sufficiently activated in the absence of cytosolic ATP production. Thus, after 30 min exposure the CP content is $14 \pm 2\%$ and the ATP content $25 \pm 2\%$ of control values (control: CP 8.7 ± 0.8 $\mu mol/g_{ww}$, ATP 6.0 ± 0.6).

When pyruvate is offered as exogenous substrate, energetic depletion is greatly delayed: after 30 min exposure CP is $45 \pm 4\%$ of control, while ATP is still $91 \pm 6\%$. That pyruvate oxidation cannot satisfy the energetic needs even in the resting muscle cells may be due to the fact that in the presence of IAA huge amounts of phosphate are trapped in the increasing stores of glycolytic intermediates. After 30 min exposure to IAA without pyruvate fructose-1,6-diphosphate has increased 115 ± 14 fold, under IAA plus pyruvate 36 ± 3 fold. Thus inorganic phosphate may become short for rephosphorylating ADP. While medium activities of GLDH and acid phosphatase remained unchanged during 60 min exposure to IAA, the time course of cytosolic enzyme release follows that of ATP decay in the presence or absence of pyruvate. The relation between enzyme release and total ATP content is identical in both cases (Fig. 2). Lactate production was negligible under IAA $\pm$ pyruvate. Thus, in these experiments differences in glycolytic energy production cannot account for differences in cytosolic enzyme release. In contrast, the actual total ATP level seems to determine the rapidity of early protein loss.

Absence of enzyme release from cell compartments other than the cytosol together with the fact that the amount of enzyme activity released at a given ATP level is even lower than in the case of anoxia, indicates that under IAA again early enzyme release is not due to cellular lysis (10). Lower enzyme release for a given total ATP level in the presence of IAA compared to anoxia might indicate that the thermodynamic utility rather than the amount of ATP present determines the extent of this gradual protein loss. Due to the trapping of inorganic phosphate in glycolytic intermediates, the free energy change of ATP hydrolysis should be higher in the presence of IAA at given ATP and ADP levels (7). Although these findings demonstrate that there is no di-

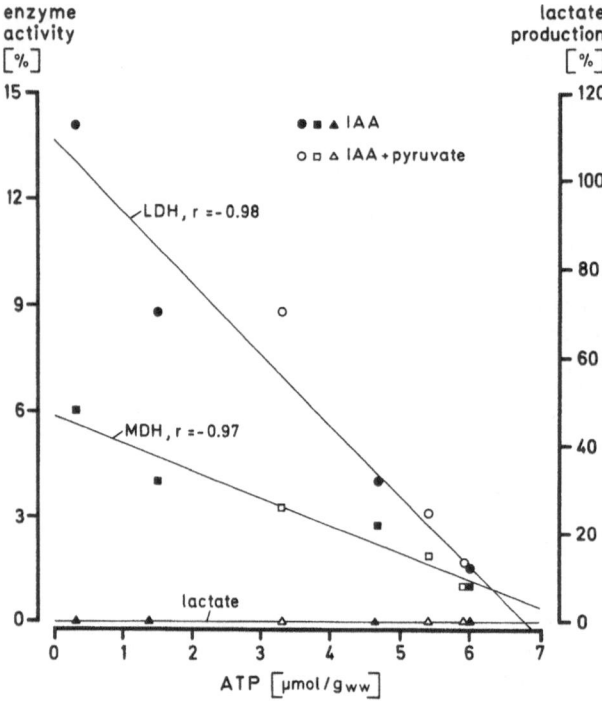

Fig. 2. Blockade of glycolysis with 1 mM iodoacetate (IAA): relation of released activities of LDH and MDH (expressed as a percentage of the initial total activity) and of lactate production rate (expressed as a percentage of 1.2 μmol lactate/g_{ww} × min) to the actual ATP content in presence or absence of 5 mM pyruvate. Each datum represents the mean value of 5 independent experiments.

rect connection between cytosolic protein release and glycolytic energy production in the isolated cell system there may be an explanation why this seems to hold in tissue. If low glycolytic energy production does indeed cause more pronounced contracture, then more single cell disruptions may result caused by the mechanical cell coupling in tissue. This, consecutively, leads to higher enzyme release.

Acknowledgments

This study was supported by the Deutsche Forschungsgemeinschaft, SFB 89 – Kardiologie Göttingen. We greatly acknowledge the skilful assistance of B. Eickhoff, H. Haacke and E. Neumeyer.

References

1. Bergmeyer HU (1974): Methoden der enzymatischen Analyse. Verlag Chemie, Weinheim
2. Bricknell OL, Daries PS, Opie LH (1981): A relationship between adenosine triphosphate, glycolysis and ischemic contracture in the isolated rat heart. J Mol Cell Cardiol 13: 941–945
3. Bricknell OL, Opie LH (1978): Effect of substrates on tissue metabolic changes in isolated rat heart during underperfusion and on release of lactate dehydrogenase and arrhythmias during reperfusion. Circ Res 43: 102–115
4. Gercken G, Hürter P (1966): Stationäre Metabolitkonzentrationen im insuffizienten Säugetierherzen nach Monojodazetat- und Natriumfluoridvergiftung. Pflügers Arch 292: 100–117

5. Hearse DJ, Chain EB (1983): Effect of glucose on enzyme release from, and recovery of, the anoxic myocardium. In: Dhalla NS (ed) Myocardial Metabolism. Recent Advances in Studies on Cardiac Structure and Metabolism, Vol 3, pp 763 – 772, University Park Press, Baltimore
6. Hohl CM, Altschuld RA, Brierley GP (1983): Effects of calcium on the permeability of isolated adult heart cells to sodium. Arch Biochem Biophys 221: 197–205
7. Kammermeier H, Schmidt P, Jüngling E (1982): Free energy change of ATP-hydrolysis: a causal factor of early hypoxic failure of the myocardium. J Mol Cell Cardiol 14: 267–277
8. Kübler W, Spieckermann PG (1970): Regulation of glycolysis in the ischemic and the anoxic myocardium. J Mol Cell Cardiol 1: 351–377
9. Piper HM, Probst I, Schwartz P, Hütter JF, Spieckermann PG (1982): Culturing of calcium stable adult cardiac myocytes. J Mol Cell Cardiol 14: 397–412
10. Piper HM, Schwartz P, Spahr R, Hütter JF, Spieckermann PG (1984): Anoxic injury of adult cardiac myocytes. Basic Res Cardiol 79, Suppl: 2: 37–42

Author's address:
H. M. Piper, Zentrum Physiologie, Universität Göttingen, Humboldtallee 23, D-3400 Göttingen (F.R.G.)

Anoxia influences the lateral diffusion of a lipid probe in the plasma membrane of isolated cardiac myocytes

S. A. E. Finch[1]), H. M. Piper[2]), P. G. Spieckermann[2]) and A. Stier[1])

Max-Planck-Institut für biophysikalische Chemie, Göttingen and [2])Zentrum Physiologie, Universität Göttingen, Göttingen (F.R.G.)

Summary

Using the technique of fluorescence photobleaching recovery we have measured the characteristics of lateral diffusion of oleylaminofluorescein (OAF) in the plasma membrane of isolated rat cardiac myocytes under normoxic and anoxic conditions. The normoxic pattern is one of slow diffusion and low recovery ($D = 1.8 \pm 0.3 \times 10^{-10} cm^2/s$, $r = 0.44 \pm 0.059$), while under anoxic conditions faster diffusion and higher recovery ($D = 2.5 \pm 0.4 \times 10^{-9} cm^2/s$, $r = 0.58 \pm 0.063$) are observed, the change proceeding via an intermediate stage with a yet faster diffusing species ($D \rangle 2.5 \times 10^{-9} cm^2/s$). The process is reversible.

We hypothesize that under normoxic conditions the lateral diffusion of OAF is hindered by the division of the cell membrane into a patchwork of more or less isolated domains by lateral and longitudinal barriers of spectrin (7, 13) which are rearranged under anoxic conditions to another pattern which permits the label greater, but still not unrestricted, freedom of movement.

Key words: fluorescence photobleaching recovery, adult cardiocytes, lateral diffusion, anoxia, spectrin

Introduction

The sarcolemma has been recognised as a structure of considerable importance in the development of ischaemic injury and it has been suggested that it is the site of the primary lesion. It is seen to undergo morphological change (11) and may be subject to damage by free radical reactions and lipid peroxidation (3, 12) which alter its properties so that it no longer maintains calcium homeostasis and water balance. Changes such as these may be expected to be reflected in the physical properties of the membrane, and to gain further insight into their magnitude and time-course we have undertaken a study of lipid diffusion in the sarcolemma of isolated cardiac myocytes under normoxic and anoxic conditions.

Methods

Adult rat cardiac myocytes were isolated as described in (10). The cells were allowed to settle on serum-coated microscope cover-slips for about 20 h, with a change of medium after 5 h. The plasma membrane was fluorescently labelled by incubating the cells with a $100\mu M$ suspension of OAF (8) in buffer (125 mM NaCl, 2.6 mM KCl, 1.2 mM KH$_2$PO$_4$, 1.2 mM MgSO$_4$, 1 mM CaCl$_2$, 10 mM glucose, 10 mM HEPES, pH 7.4) for 30 min at 37 °C. After washing the excess label off, the cover-slips bearing the cells were built into a chamber in which the cells could be exposed to various media. The temperature throughout the experiment was maintained at 37 °C. Anoxia was normally induced by perifusing the cells with a buffer containing 25 U. glucose oxidase and 1000 U. catalase/ml, but for control experiments anoxic buffer containing 2.5 and 100 U./ml and buffers rendered substantially O$_2$-free by extensive bubbling with Ar or CO were also used.

The lateral diffusion coefficient of the OAF in the plasma membrane was determined by the fluorescence photobleaching recovery method (1) with an apparatus very similar to that described in (6). The data were analysed using the curve fitting procedure described in (16).

Results

Under normoxic conditions the lateral diffusion coefficient determined for OAF in the plasma membrane, $1.8 \pm 0.3 \times 10^{-10} \text{cm}^2/\text{s}$ (mean $\pm$ s.e.m.), is very low and in the range usually associated with gel phase membranes (14) (Fig. 1a). When the cells are perifused with an anoxic buffer two distinct curve forms are seen sequentially. The first is characterised by a fast-diffusing component ($D > 2.5 \times 10^{-9} \text{cm}^2/\text{s}$). Since this curve has a distinct discontinuity (Fig. 1b) it is probable that it is caused by membrane flow (1). After some time it is replaced by another curve-type (Fig. 1c) which gives a diffusion coefficient of $2.5 \pm 0.4 \times 10^{-9} \text{cm}^2/\text{s}$ on evalution. The normoxic and anoxic fluorescence recovery curves are distinguishable by two other characteristrics: (a) the recovery is greater when the cells are anoxic – it rises from 0.44 ± 0.059 when the cells are perifused by a normoxic buffer to 0.58 ± 0.063 and (b) more label is bleached for a given light pulse – if this is adjusted for 50% bleach under normoxic conditions, a 70% bleach is observed when the cells become anoxic.

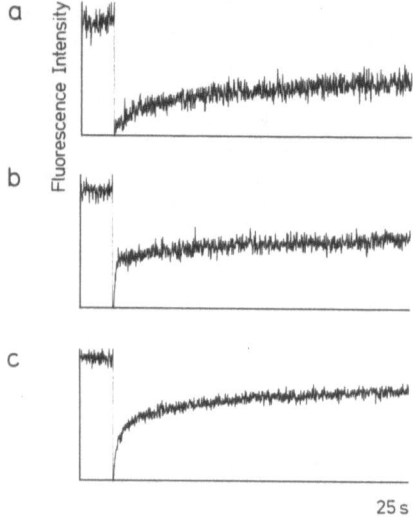

Fig. 1. FPR curves showing the changing pattern of lateral diffusion of OAF in the sarcolemma of (a) normoxic, (b) intermediate and (c) anoxic cardiac myocytes.

When anoxic cells are again perifused with a normoxic buffer the lateral diffusion characteristics revert to the normoxic pattern, and if anoxic buffer is once again perfused, the diffusion characteristics also change. These alterations take place without any morphological change apparent under the light microscope at 1000-fold magnification. The time-course of each experiment is dependent on the O_2 tension of the buffer used. At its fastest the change from normoxic to anoxic diffusion pattern can be complete about 3 min after the start of perifusion with anoxic buffer and is probably limited by the kinetics of removal of O_2 within the sample chamber. The reverse process has similar characteristics.

To investigate the possibility that the changes described might be due to the reaction of $^1O_2$ produced by the quenching of triplet fluorescein in the bleach/monitoring spot with membrane components we did several experiments in which either 1 mM 1,4-diazabicyclo (2.2.2)-octane – an oxygen scavenger (5, 15) – was included in the buffer, or the cells were perifused with a retinol suspension (5). Neither treatment had any effect on the course of the experiment. The fact that the change from normoxic to anoxic diffusion pattern proceeds via an intermediate stage and is not gradual with falling O_2 tension, but sudden, is also evidence that activated oxygen species are not the sole cause of the phenomena observed. Finally, in a control experiment anoxia was found to have no effect on the diffusion on OAF in the hepatocyte plasma membrane.

Discussion

While most of the figures obtained from the experiments are unremarkable – incomplete fluorescence recovery is the rule in natural membranes (4, 9) and the lateral diffusion coefficient of $2.5 \times 10^{-9} cm^2/s$ under anoxic conditions also lies in the expected range (9) – the value obtained under normoxic conditions is extraordinarily low and requires explanation.

The organisation of the cardiac myocyte plasma membrane is notable for its periodicity: the entrances to the T-tubuli lie every 2 μm or so and the whole overlies a spectrin framework comprising transverse members associated with the M lines and Z discs and longitudinal ones between the myofibrils (7, 13). The protein scaffold so formed has a rectangular pattern with a side length in the μm range and is connected to membrane-bound anion-transport proteins via ankyrin (2). The bleach/monitoring spot used in this study had a diameter of 1.2 μm, so the measurements may well be sensitive to any changes in this organisation. The data presented here are consistent with the idea that the spectrin network and associated proteins form more or less lipid-impermeable barriers in the membrane with the consequence that both fluorescence recovery and diffusion coefficient are small under normoxic conditions.

When O_2 tension falls below a critical level this network undergoes a rearrangement – it would seem that one structure is first demolished, hence the high diffusion transiently observed – to be replaced by another which permits the lipid to diffuse over greater distances. The relation between the effect on lateral diffusion and the formation of blebs reported in (11) is uncertain. The diffusion coefficient is constant over the whole cell surface, while blebs are first formed at the poles of the cell (11).

References

1. Axelrod D, Koppel DE, Schlessinger J, Elson E, Webb WW (1976): Mobility measurement by analysis of fluorescence photobleaching recovery kinetics. Biophys J 16: 1055–1069
2. Branton D, Cohen CM, Tyler J (1981): Interaction of cytoskeletal proteins on the human erythrocyte membrane. Cell 24: 24–32
3. Buja LM, Chien KR, Burton KP, Hagler HK, Mukherjee A, Willerson JT (1983): Membrane damage in ischemia. In: Myocardial injury; Spitzer JJ (ed). Plenum Press, New York, 421–431
4. Cherry RJ (1979): Rotational and lateral diffusion of membrane proteins. Biochim Biophys Acta 559: 289–327
5. Foote CS (1976): Photosensitized oxidation and singlet oxygen: Consequences in biological systems. In: Free radicals in biology, Vol 2; Pryor WA (ed). Academic Press, New York, 85–133
6. Kapitza H-G, Sackmann E (1980): Local measurement of lateral motion in erythrocyte membranes by photobleaching technique. Biochim Biophys Acta 595: 56–64
7. Nelson WJ, Lazarides E (1983): Expression of the β subunit of spectrin in nonerythroid cells. Proc Natl Acad Sci USA 80: 363–367
8. Organikum: Organisch-Chemisches Grundpraktikum (1965): VEB Deutscher Verlag der Wissenschaften, Berlin, 396–397
9. Peters R (1981): Translational diffusion in the plasma membrane of single cells as studied by fluorescence microphotolysis. Cell Biol Int Reports 5: 733–760
10. Piper HM, Probst I, Schwartz P, Huetter JF, Spieckermann PG (1982): Culturing of calcium stable adult cardiac myocytes. J Mol Cell Cardiol 14: 397–412
11. Piper HM, Schwartz P, Spahr R, Huetter JF, Spieckermann PG (1984):Early enzyme release from myocardial cells is not due to irreversible cell damage. J Mol Cell Cardiol 16: 385–388

12. Rao PS, Mueller HS (1983): Lipid peroxidation and acute myocardial ischemia. In: Myocardial injury; Spitzer JJ (ed). Plenum Press, New York, 347–363
13. Repasky EA, Granger BL, Lazarides E (1982): Widespread occurrence of avian spectrin in nonerythroid cells. Cell 29: 821–833
14. Vaz WLC, Derzko ZI, Jacobson KA (1982): Photobleaching measurements of the lateral diffusion of lipids and proteins in artificial phospholipid bilayer membranes. In: Membrane reconstitution; Poste G and Nicolson GL (eds). Elsevier Biomedical Press, Amsterdam, 83–136
15. Wassermann HH, Murray RW (eds) (1979): Singlet oxygen. Academic Press, New York
16. Yguerabide J, Schmidt JA, Yguerabide EE (1982): Lateral mobility in membranes as detected by fluorescence photobleaching recovery. Biophys J 39: 69–75

Author's address:

Dr. S. A. E. Finch, Max-Planck-Institut f. biophys. Chemie, Am Faßberg, D-3400 Göttingen (F.R.G.)

Summaries of the discussions of the symposium

J. Y. Cheung[1]), B. Koidl[2]), H. M. Piper[3]) and G. Trube[4])

Renal Unit, Massachusetts General Hospital, Boston, (U.S.A.)[1])
Institut für Medizinische Physik, Universität Graz, Graz (Austria)[2])
Zentrum Physiologie, Abt. Herzstoffwechsel, Universität Göttingen, Göttingen (F.R.G.)[3])
Max-Planck-Institut für biophysikalische Chemie, Göttingen (F.R.G.)[4])

Foreword

In this chapter we attempt to give some reflection of the discussions during the symposium. Since discussions by themselves are already selective and only some of the points raised can be represented here, these summaries must be to some extent an arbitrary collection of arguments. However, not only for beginners, but also for experts in the art of isolating myocytes, it might be helpful to hear some of these informal suggestions and criticisms. It may turn out that some of the arguments become quickly out of date. But this would be a promising perspective, since the state of the art still urgently demands very basic investigations. It certainly will favour a rapid further development of the myocyte model, when experts of all the technical disciplines interested in this model meet again and communicate, in spite of differences in their specific interest.

1. Cell isolation

(Discussions raised by papers of Powell [pp. 9–12], Bendukidze et. al. [pp. 13–18], Bechem et al. [pp. 19–22])

In the discussion regarding the different techniques of cell isolation, the consensus was that current methods are entirely empirical and based on trial and error since the mechanism of cell separation is not fully understood. Because dissociation of heart tissue by highly purified collagenase alone is impossible, the contaminating proteases in crude collagenase preparations seem important for the separation of the cells. It was generally felt that no relation could be found between the collagenase activity of various batches and the success in isolating single cells.

As pointed out by Eckel (2) trypsin may be helpful to digest damaged round cells and hence facilitates their removal from the cell suspension. Experience with pronase for dissociation of cells is limited. The results presented by Isenberg (pp. 13–18) showed a normal electrophysiological behaviour of cells dissociated by pronase. Additionally, it was reported by Isenberg that there is no indication that Na^+ channels are affected, since the speed of upstroke of the action potential is not changed when cells are isolated by pronase.

For storage of cells in suspension it was noted that vigorous oxygenation is not necessary. Exposure to air is sufficient for adequate oxygen supply if dense packing of cells is avoided. Isolated cardioballs seem especially useful for experiments investigating the effects of changing the intracellular milieu on membrane currents. Due to their small size the cytoplasmic medium can be exchanged within 5 min, when using a suction pipette both for electrical recording and internal dialysis. The relatively rapid morphological changes observed in atrial cell cultures are apparently specific for that cell type, since ventricular cells cultured in an identical manner do not exhibit the same behaviour (5).

2. Cell viability and cell surface properties

(Discussions raised by papers of Cheung et al. [pp. 23–29], Borgers et al. [pp. 31–36], Piper et al. [pp. 37–42], Bailey et al. [pp. 43–46], Maisch [pp. 47–52])

It was stated in the presentation of Cheung (pp. 23–29) that rod-shaped morphology is a sufficient criterion of cell viability. However, it has been observed in electrophysiological laboratories that a minority of isolated cells, while maintaining intact morphology and exhibiting normal electrophysiological properties, failed to contract on electrical stimulation.

The validity of succinate-stimulated oxygen consumption as a criterion of cell viability, proposed by Farmer et al. (3), was discussed. It was stated that in calcium-tolerant cells the proportion of damaged cells may be estimated from the enhancement of oxygen consumption by addition of succinate. However, this presupposes that damaged myocytes while being permeable to succinate still retain normal respiratory capacity. That this may not be generally the case is indicated by the reports of other laboratories in which no correlation between percent of damaged cells and the degree of succinate-stimulated oxygen consumption was found.

A good indication of impending cell death appears to be formation of blebs observed at the light microscopic level. This formation of rather large blebs is always associated with cell contracture. Bleb formation is often preceded by a granular appearance of the myocyte – which may represent either mitochondrial swelling or formation of microblebs on the sarcolemma detectable only by electron microscopy. It was stated that microbleb formation is reversible on reoxygenation and it was hypothesized that these protrusions of the μm-range may arise from localized alterations in the subsarcolemmal cytoskeleton.

It was not clear whether development of contracture in anoxic isolated myocytes is secondary to loss of cellular ATP or to rise in cytosolic calcium. Current evidence does not allow clear delineation of the mechanism of contracture. It was stated that elevation in total cell calcium as measured by atomic absorption is not necessarily related to cell death; conversely, cell death may occur without a rise in total cell calcium.

The histochemical method described by Borgers (pp. 31–36) seems to give a quantitative demonstration of intracellular calcium localization. The method's threshold of detection for calcium is not known although it was stated that below 10^{-5} M calcium precipitates of calcium complexes were not observed. Accumulation of calcium precipitates near the sarcolemma is thought to be due to calcium complexing with acidic phospholipids. The sarcolemmal precipitates can be displaced by lanthanum and low calcium perfusion and are enhanced by positive inotropic agents. It is at present not known why subsarcolemmal calcium precipitates are not present in severely ischemic cardiac cells.

The novel technique of condensed phase radioluminescence (CPR; Bailey et al., pp. 43–46) and the antimyolemmal antibody staining technique (AMLA; Maisch, pp. 47–52) take advantage of the direct sarcolemmal access in an isolated-myocyte system. The CPR technique seems to promise elimination of non-specific binding effects in that non-specific binding will not result in adequate energy transfer to provide a detectable light signal. In addition, CPR allows one to measure binding of drugs and its effect on cell metabolism concurrently since destruction of cells is not necessary for CPR. This technique has been validated in the case of β-blockers since results obtained from intact cells and isolated membranes are comparable. The AMLA technique was widely acknowledged to hold great promise in facilitating the clinical diagnosis of viral-induced myocarditis, even if the exact nature of the antibody and its target antigen remains unknown at present.

3. Metabolism and sarcolemmal transport properties

(Discussions raised by papers of Spahr et al. [pp. 53–56], Kammermeier et al. [pp. 57–60], Eckel et al. [pp. 61–64], Rauch et al. [pp. 65–68])

Concern was expressed regarding the high lactate output in isolated myocytes. It was pointed out by Probst (pp. 53–56) that increasing oxygen tension 5-fold in the incubation medium (oxygen vs. air) did not result in a reduction of lactate output. In addition, dichloroacetate (DCA) which stimulates pyruvate dehydrogenase was able to increase CO_2 production, indicating oxygen availability was sufficient. Furthermore, it was reported that while liver *in vivo* does not produce net lactate, isolated hepatocytes in culture also produce measurable amounts of lactate despite adequate oxygenation (6). These observations suggest that lactate production is a general finding in cultured cells, perhaps due to pyruvate dehydrogenase inactivation.

The kinetics of lactate transfer over the sarcolemma seem to support the assumption of a carrier mechanism for L-lactate. While lactate uptake exhibits sigmoidal saturation kinetics, pyruvate inhibits lactate uptake at lower lactate concentrations and enhances it at very high lactate concentrations (< 25 mM). The physiological significance of this interaction is at present not clear.

The cell isolation technique as employed by Eckel et al. (pp. 61–64) raised some questions regarding the sensitivity of insulin receptors to trypsinisation. It was pointed out by Eckel that trypsin at the applied concentration and duration of exposure probably reduces the number of receptors by 10%. Prolonged exposure to the enzyme results in complete removal of the insulin receptor. Measurements at one hour after isolation were reported to be similar to those obtained at one day in culture. Loss of ATP due to EDTA treatment was raised as a possible explanation for the difference in insulin binding and 3-O-methylglucose uptake experiments. But it was stated that detectable loss of ATP was not observed during 15 min of EDTA treatment, a time sufficient to perform the experiments.

The use of thallium scintillography in the clinical setting to estimate myocardial tissue viability may represent an overestimation since experiments employing isolated cells demonstrated that permeable cells may still have appreciable thallium uptake. On the other hand, the use of labelled palmitate in determining metabolically active myocardium *in vivo* is critically dependent on substrate transfer across the endothelial wall. This may result in an underestimation of viable tissue. Thus, neither tracer method gives fully reliable indications of tissue viability. However, at the precision necessary in clinical practice, they still may be valuable tools.

4. Culture conditions

(Discussions raised by papers of Lundgren et al. [pp. 69–74], Schwartz et al. [pp. 75–78], Jacobson et al. [pp. 79–82], Spahr et al. [pp. 83–86])

In the discussion regarding cell attachment factors, it was mentioned that adult cells may attach on certain petri dishes without any further treatment (e. g. 'Permanox' from Lux). However, since commercially available culture dishes vary in quality from batch to batch, cell attachment without further treatment is often suboptimal. It was recommended that rinsing culture dishes with saline be routinely performed prior to plating of cells. Culture media *per se* do not appear to affect attachment. Similarly, albumin provides little to no enhancement of cell attachment. Raising the temperature of incubation medium partially reduces cell attachment.

The question of how cells dissociate from each other during the separation process was again raised. While it is generally known that on cell separation the entire gap junction complex can be observed within the cytoplasm of one cell, it remains unclear whether splitting of gap junctions can occur. If gap junction complexes are transferred *in toto* from one cell to the other, than the mechanism by which cells repair the membrane defect is unknown.

On account of the impressive morphological changes of the cultured cells as described by Jacobson et al. (pp. 79–82), concern was expressed regarding the relevance of cultured cells to the adult heart *in vivo*. It was pointed out that after an intermediate stage of development, cultured cells regain certain properties of the adult cell, namely, normal resting membrane potentials, action potential initiated by a fast sodium current, and re-organization of myofilaments in localized regions of a cell. It was reported (Jacobson) that if cells were isolated from rats at least 3 weeks old and then placed in culture, cell divisions were not observed, the number of nuclei per cell remained constant, and the rate of DNA synthesis remained low. It was felt that while long-term cultured cells may not resemble adult myocardium in all respects, it remains a useful tool to investigate the development potential of a supposedly 'terminally differentiated' cell type, and to study cell-cell interaction *in vitro*.

While it is not known by which mechanism hepatocytes influence the development of isolated myocytes in culture, the presence of a hepatocyte monolayer enables cultured myocytes, attached on this substratum, to progress directly from the rod-shaped morphology to the spread-out configuration without the intervening rounded stage as observed in the culture system described by Jacobson. It was speculated that the intermediate round stage may be a consequence of delayed cell attachment.

5. Electrophysiology

(Discussions raised by papers of Powell (pp. 87–92), van der Heyden et al. (pp. 93–96), Trube (pp. 97–100), Bechem et al. (pp. 101–106), Osterrieder et al. (pp. 107–110))

Single cells are superior to multicellular preparations for voltage-clamp experiments, since problems caused by series resistances and accumulation of ions in a narrow intercellular space are avoided. However, similar difficulties could still arise from the presence of the T-tubules, although at a reduced scale.

It was mentioned that when the K^+ conductance is blocked (e.g. by Cs^+) cells depolarize, and that the subsequent contraction can be prevented by removing Ca^{2+}. Then, substitution of Ca^{2++} by Mg^{2+} may ameliorate the harmful effects of low Ca^{2+} solutions.

It may be impossible to clamp the membrane potential of ventricular cells to positive potentials at a strongly increased potassium conductance as, e.g., after metabolic inhibition by KCN. Therefore, it was suggested that atrial cardioballs are better suited for these experiments. The K^+-outward current of cardioballs is increased when the cells are dialysed internally by a solution without ATP (a condition which may be similar to metabolic inhibition). This increase is mediated by the K^+ channels activated by the lack of ATP. However, in spite of qualitative agreement of the ventricular and the atrial cell model, there is some difference between the current-voltage relations. It is not clear whether this is due to a difference between the methods or the preparations. The biological relevance of increased K^+ conductance during energy depletion states such as ischemia is that, due to the shortening of the action potential, contraction is reduced, and consecutively energy consumption is diminished.

It was pointed out that after ATP depletion by KCN treatment addition of glucose may re-establish normal potassium conductance and normal action potential duration despite persistently lower than normal ATP levels. The apparent ability of low ATP levels to maintain normal membrane electrophysiological parameters may be explained by the fact that ADP and inorganic phosphate levels are also low, thus the available free energy from ATP hydrolysis may be close to normal values.

6. Cell contraction

(Discussions raised by papers of Koidl et al. (pp. 111–116), Isenberg et al. (pp. 117–122), Piper et al. (pp. 123–128), Rappaport et al. (pp. 129–132))

It was mentioned that a resting sarcomere length of ~ 1.9 μm measured on isolated cells represents the true slack length of a cardiac cell. *In situ,* however, the sarcomere length during diastole is longer (~ 2.1 μm), since the cells are slightly stretched. The minimum sarcomere length (1.5 μm) attained during contraction in an isolated myocyte (Isenberg) is shorter than that found in the organ. At this sarcomere length, considerable distortion of myosin filaments may occur and this may account for the faster relaxation of isolated myocytes.

During an action potential of several 100 ms duration the non-inactivating part of I_{si} may directly support the contraction (Isenberg). The communications by Isenberg et al. (pp. 117–122) and Bechem et al. (pp. 101–106) similarly showed that a substantial part of I_{si} itself is small. However, no conclusion could be reached whether this shows that I_{si} is inactivated by an increase of intracellular Ca^{2+} or by a voltage-dependent gating process, which may be more complicated than assumed in Hodgin-Huxley models. For isolated single cells, the question whether there is mechanical activation close to the threshold of I_{Na} was denied. Earlier reports about this phenomenon in multicellular preparations could be due to an escape of the membrane voltage during the activation of I_{Na}.

Concerning the relevance of the temperature-dependent effects of calcium antagonists on the maximum rate of pacing, it was mentioned that the efficiency of calcium antagonists during cardioplegia may similarly be reduced. It was pointed out (Grützmann) that in an isolated perfused working rat heart preparation, the relative decrease in aortic systolic pressure on exposure to nifedipine was similar whether at 37 °C or 22 °C. It should be made clear that in the isolated cell study only the effect of nifedipine on the excitaion of contractions was investigated (4). This is in contrast to the intact organ in which chronotropic, inotropic, and electromechanical coupling effects of the drug all act in concert to determine the final cardiac output. It is very difficult, if not impossible, to discern the temperature dependence of the different effects of nifedipine on the heart based on *in vitro* organ perfusion studies.

The structure proteins responsible for the stable rod-shape of isolated ventricular cells remains unknown. Desmin and tubulin do not seem necessary for conserving the shape, because for example tubulin may undergo considerable reorganisation during hypertrophy (Rappaport et al., pp. 129–132).

7. Oxygen damage and anoxic injury

(Discussions raised by papers of Noronha-Dutra et al. (pp. 133–136), Van der Laarse et al. (pp. 137–142), Piper et al. (pp. 143–148), Finch et al. (pp. 149–152)

Although the morphological changes observed in isolated myocytes incubated with serum obtained from isoprenaline-treated rats and those incubated in the presence of H_2O_2 are similar, it remains to be established that damage by isoprenaline-treated serum is mediated by oxygen free radicals. The source of the free radicals supposedly generated by isoprenaline treatment is at present unknown. Unlike adult heart cells in suspension, neonatal cell monolayer culture systems permit the evalutation of the role of cell-cell interactions in anoxia-induced damage. Mechanical forces from cell-cell interaction may explain the differences of damage on reoxygenation between neonatal cell monolayer and the adult heart cell culture (Piper et al., pp. 37–42 and pp. 143–148). The qualitative relationship between percent of cell death and LDH release may be explained by the hypothesis that superimposition of mechanical forces from neighbouring cells on cell membranes already damaged by anoxia results in cell lysis and thus release of the entire LDH content of the cell. It was debated that *in vivo*, enzyme release as observed in muscle exercise may not indicate cell death as suggested by the above finding. The relatively long incubation time in anoxia to induce measurable cell damage raised the question whether neonatal cell cultures are indeed relevant models for the study of ischemic heart disease.

There has been a suggestion from studies using isolated perfused hearts that depletion of ATP from glycolytic sources is more reponsible for cell contracture than depletion of ATP from oxidative pathways (1). However, it was pointed out that compartmentation of ATP should be interpreted not as discrete anatomical entities but in a functional sense, on the basis of differences in available free energies for hydrolysis of ATP derived from glycolytic and oxidative sources (Kammermeier).

While evidence from fluorescence microphotolysis studies shows that membrane fluidity increases with anoxia, it is at present not known how this is brought about, e.g., change in conformation of lipids or proteins or both. Caution must be exercised regarding the intensity of light incident on the probe, since at high light intensities photochemical reactions may occur. In the experiments described (Finch et al., pp. 149–152), reducing light intensity essentially gave the same results except for a lower signal-to-noise ratio. On the other hand, increasing light intensity apparently led to formations of cross-links between the probe and membrane components as suggested by the relative immobility of the fluorescent probe.

8. Pharmacological screening

(Discussions raised by introductory remarks from I. Bailey, Macclesfield, M. Borgers, Beerse, R. Grützmann, Wuppertal)

The advantage of using isolated adult myocytes as a model for pharmacological screening is manifold:

(I)   isolated myocytes provide a well-defined experimental system in which the extracellular environment can be controlled precisely;

(II)   compared to intact organ perfusion models or whole animal studies, the effects of a large variety of drugs can be studied on a single preparation, thus making this model cost-effective and allowing a reduction in animal use in drug screening;

(III)   reproducibility of experimental results is superior to models using intact organs because isolated cell systems circumvent heterogeneity of cell types, coronary perfusion differences, and the uncertainty of transfer of substrates/drugs from the intravascular compartment to the cell;

(IV)   the model provides a unique system for studies of drug/hormone binding to receptors on a living cell as compared to isolated cell membranes;

(V)   it allows direct visualization and access to the vital myocyte for electrophysiological and microspectrophotometric studies;

(VI)   unlike embryonic cell culture systems which have been in widespread use, isolated adult myocytes retain most of the properties of the intact ventricular myocardium and thus results are more relevant to the *in vivo* heart ventricle of adult animals.

However, one must bear in mind that while studies using single cells provide unambiguous results on the metabolic, electrophysiological and morphological effects of various drugs on the myocardial cell, it is difficult to extrapolate the results to the whole heart *in vitro* since different cell types, e.g. SA-nodal and ventricular cells, may respond differently to the drug. Therefore, future attempts should be directed to improving isolation techniques for different cell types of the myocardium so that the effect of a drug can be studied in each type. Presently, embryonic cell cultures may be the best electrophysiological single-cell model mimicking the sinus node for detection of possible drug effects. In organs *in vivo*, neural, circulatory, and hormonal influences may play a role in forming the integrated response of the animal to a drug, which, of course, cannot be predicted from single-cell studies. In addition, no information can be obtained from isolated myocyte investigations on drug toxicity in other organs.

In summary, isolated myocytes provide a useful model for initial drug screening studies in that a large number of drugs can be examined in a relatively rapid fashion. Drugs that have shown significant effects on isolated cells can then be subjected to more integrative studies using isolated perfused organs and whole animals.

References

1. Bricknell OL, Daries PS, Opie LH (1981): A relationship between adenosine triphosphate, glycolysis and ischemic contracture in the isolated rat heart. J Mol Cell Cardiol 13: 941–945
2. Eckel J, Pandalis G, Reinauer H (1983): Insulin action on the glucose transport system in isolated cardiocytes from adult rat. Biochem J 212: 385–392
3. Farmer BB, Harris RA, Jolly WW, Hathaway DE, Katzberg A, Watanabe AM, Whitlow DR, Besch HR (1977): Isolation and characterization of adult rat heart cells. Arch Biochem Biophys 179: 545–558
4. Piper HM, Hütter JF, Spieckermann PG (1984): Temperature dependence of nifedipine action. J Mol Cell Cardiol 16: 277–280
5. Piper HM, Probst I, Schwartz P, Hütter JF, Spieckermann PG (1982): Culturing of calcium stable adult cardiac myocytes. J Mol Cell Cardiol 14: 397–412
6. Wölfle D, Schmidt H, Jungermann K (1983): Short-term modulation of glycogen metabolism, glycolysis and gluconeogenesis by physiological oxygen concentrations in hepatocyte cultures. Eur J Biochem 135: 405–412

Author's address:

Dr. med. H. M. Piper, Zentrum Physiologie, Abt. Herzstoffwechsel, Humboldtallee 23, D-3400 Göttingen (F.R.G.)

Abstracts of the poster-session on the adult myocyte model

Isolation of adult ventricular myocytes for electrophysiological experiments. C. Achenbach, J. Wiemer and R. Preisler (Dept. of Physiology, Wilhelmstr. 31, D-5300 Bonn, F.R.G.)

Ca-tolerant adult cardiac myocytes can be prepared by enzymatic digestion in Ca-free solutions containing high concentrations of potassium-L-glutamate (130 μM) and very low sodium. During the separation and hyperpermeability of the cells, this effectively prevents sodium overload. Guinea-pig Langendorff preparations were perfused with trypsin (1‰ w/v) for 10 min preceding 10–15 min collagenase (1‰ w/v) plus protease (0.02‰ w/v). After this treatment in high potassium, the hearts were immediately transferred to normal Tyrode solution (1.8 mM Ca). Initially, the proportion of rod-shaped cells was near 40% ($37 \pm 6\%$, n = 13). Rods could be enriched by sedimentation or re-incubation in solutions of normal Tyrode containing enzymes. In 1.8 mM Ca, the half-lives of rod populations were 7–8 hours and the cells remained quiescent when irritated mechanically or exposed to temperature changes. This behaviour, and their electrical parameters (resting potential: -81 ± 6 mV; overshoot: 40 ± 10 mV;; rate of rise: 300 V/s; APD: between 300–400 ms) showed ventricular cells isolated in this manner to be well-suited for lengthy electrophysiological experiments.

Pharmacological protection against K^+-induced Ca^{2+}-overload in cardiac myocytes. M. Borgers and F. Thoné (Laboratory of Cell Biology, Department of Life Sciences, Janssen Pharmaceutica, Beerse, Belgium)

Myocytes were isolated from adult rat hearts and made Ca^{2+}-tolerant according to Piper et al. (J Mol Cell Cardiol 14: 397–412, 1982). Quiescent Ca^{2+}-tolerant cells adhering firmly to the bottom of petri-dishes were all rod-shaped. The number of cells in a well delineated area of the disk were scored. Then, Ca^{2+}-overload was induced by increasing the K^+ concentration of the Krebs-Henseleit medium to 50 mM. After 30 min incubation at 30° C the number of rod-shaped myocytes was less than 10%. Ca^{2+}-intolerant cells were severely contracted and frequently detached from the dish. Ultrastructurally, such cells demonstrated severe intracellular Ca^{2+}-overload. Protection against Ca^{2+}-intolerance was assessed by adding different Ca^{2+}-entry blockers in the $10^{-7}–10^{-5}$M range to the medium 15 min prior to exposure to high K^+. Protection ranged between 80 and 20% and the order of potency was lidoflazine > flunarizine > nifedipine > verapamil. These results suggest that potency to prevent K^+ induced Ca^{2+}-overload is not correlated with slow channel blockade since lidoflazine and flunarizine are practically devoid of slow channel blocking properties.

Isolation of single cardiac Purkinje cells. G. Callewaert, E. Carmeliet and J. Vereecke (Laboratory of Physiology, University of Leuven, Gasthuisberg, B-3000 Leuven, Belgium)

For studies on membrane excitability, single cell preparations have distinct advantages over the conventional multicellular preparations. While several methods have been reported for the isolation of single myocytes, the isolation of single Purkinje cells is hampered by the fact that these cells are tightly packed and surrounded by a thick connective tissue sheath.

We have developed a technique which provides viable single Purkinje cells. Purkinje fibres are injected with a Ca-free collagenase containing Tyrode solution into the space between cells and connective tissue sheath.

By this procedure single cells are obtained. A fraction of these cells appear normal under the light microscope and possess electrophysiological properties comparable to those reported for Purkinje fibres. The action potentials have normal configurations and the cells respond to elevated K, TTX and adrenaline in the same way as the multicellular preparation. Under voltage clamp conditions, a pacemaker current can be recorded which behaves as an inward current activated upon hyperpolarization.

These findings indicate that the isolated cells are suitable for electrophysiological studies of the cardiac Purkinje system.

Calcium homeostasis in isolated cardiac myocytes. J. Y. Cheung and J. V. Bonventre (Departments of Medicine, Massachusetts General Hospital and Harvard Medical School, Boston, MA 02114, U.S.A.)

Cytosolic calcium activity in isolated, calcium-tolerant myocytes has been measured using the Null-point Titration Method with Arsenazo III as the extracellular free calcium indicator. Briefly, myocytes are suspended in buffer containing varying concentrations of free calcium. After a steady Arsenazo III signal is obtained, digitonin is added to selectively render the plasma membrane permeable. Depending on the initial extracellular calcium activity, an increase (efflux of cell calcium) or decrease (influx of extracellular calcium) in Arsenazo III signal can be observed. The cytosolic free calcium is determined by extrapolating to the extracellular calcium activity at which there is no net influx or efflux of calcium – the Null-point. Under control conditions, cytosolic calcium activity was 0.32 ± 0.03 μM (n = 9). When succinate (2 mM) was present in the medium, cytosolic free calcium decreased to 0.15 ± 0.02 μM (n = 5, p < 0.02) suggesting that the set point of cytoplasmic free Ca^{2+} may be determined by the respiratory rate of the mitochondria. Neither vasopressin (1 μM) nor norepinephrine (1 μM) have any appreciable effect on cytosolic free calcium concentrations, the values being 0.35 ± 0.10 μM, and 0.34 ± 0.10 μM, respectively. We are at present studying the role of the major intracellular storage sites for calcium in the regulation of cell calcium activity in the isolated myocardial cell.

Adult cardiac myocytes in primary culture: a new model for studies on heart metabolism. G. van Echten, J. Eckel and H. Reinauer (Diabetes Research Institute, Düsseldorf, F.R.G.)

An *in vitro* system for long term studies on adult heart metabolism is presently lacking. Therefore, attempts have been made to establish a primary culture of adult cardiac myocytes, isolated by collagenase perfusion of rat hearts. Using serum-free BM-86 Wissler medium the cells could be maintained in monolayer for at least three weeks. Up to 10 days the cells exhibited a rod shaped morphology, unaltered ultrastructure and a constant level of ATP (43.5 nmol/mg protein) and ADP (4.2 nmol/mg protein). During the third week in culture a process of dedifferentiation was observed, resulting in irregularly shaped cells.

Studies on insulin binding demonstrated the presence of specific insulin receptors, with a constant number and affinity up to at least 3 days. In the presence of insulin (10^{-11} mol/l) a constant rate of protein synthesis was measured up to 63 hours in culture. When compared to control cells cultured in the absence of insulin, an increasing stimulation of protein synthesis by insulin (1.7×10^{-11} mol/l and 1.7×10^{-7} mol/l) was observed dependent on time in culture reaching a value of 35 and 107%, respectively, after 63 hours in culture.

In light of the retention of several metabolic functions, hormone receptors and hormone responsiveness, primary cultured cardiac myocytes may be regarded as a suitable model for long-term studies on cardiac metabolism.

Enhanced cellular Na load on isolated cardiac myocytes during action potentials prolonged by ATX II. G. Isenberg and U. Ravens (II. Physiologisches Institut, 6650 Homburg/Saar and Abteilung Pharmakologie der Universität, 2300 Kiel, F.R.G.)

In myocytes isolated from bovine or guinea-pig ventricles the Anemonia sulcata toxin ATX II enhances the extent of shortening and prolongs the action potential (AP) duration. Using a single patch electrode (Hamill et al., 1981, Pflügers Arch 391:85) net membrane currents in response to long lasting (8.4 s) voltage clamp steps were recorded. 20 nM ATX II induces a slowly decaying, TTX-sensitive inward current which is responsible for the prolonged AP. We conclude that ATX II modifies Na channels by retardation of their inactivation. ATX II-induced i_{Na} is obtained by subtracting the current traces in the presence and absence of ATX II. Provided that Na ions are the main charge carrier during the ATX II-induced current and provided that Na ions distribute evenly throughout the cell volume, the increase in intracellular Na concentration during an 8.4 s clamp step from -80 to -10 mV is calculated to be 0.7 mM. Extrapolating this Na load to an 1 s long AP yields a concentration increase of 0.16 mM Na, which is about 4 times larger than the amount of sodium entering the cells during unmodified i_{Na} (Brown et al., 1981, J Physiol 318:479). The shift in apparent reversal potential of the ATX II-induced current during an 8.4 s clamp step also suggests transient Na accumulation.

An opto-electronical system for analysing contraction patterns of single myocardial cells. R. Koch and H. M. Piper (Tierärztliche Hochschule Hannover, Abt. für Medizinische Physik, Bischofsholer Damm 15, D-3000 Hannover and Zentrum Physiologie, Universität Göttingen, Humboldtallee 23, D-3400 Göttingen, F.R.G.)

For optical recording of the isotonic contractions of isolated adult heart cells a simple electronic optodetector system was developed. It can be attached to standard microscopes. The picture of the cell is focused on a photo element producing an electrical current proportional to the light intensity. An analog circuit converts the current into a voltage. With cell contraction the light intensity is modulated resulting in a modulated electric signal (u[t]). This signal is amplified and noise filtered and can be used for other analog or digital signal-processings. An analog system calculates the temporal derivative to obtain a signal proportional to the speed of contraction (du[t]/dt). Other processed signals are instant frequency or interval time and the amplitude of contraction. In a stimulator unit, pulses of variable duration, interval and amplitude are generated for electrical stimulation of the cells. All signals may be recorded by a multichannel thermal paper-recorder.

Does β-adrenergic stimulation increase the number of functional Ca-channels in the heart?[1]). W. Osterrieder, G. Brum (II. Physiologisches Institut der Universität des Saarlandes, D-6650 Homburg/Saar, F.R.G.)

It has been shown that adrenaline, cAMP and the catalytic sub-unit of cAMP-dependent protein kinase increase the slow inward Ca current in the heart (Brum et al., Pflügers Arch 398: 147–157, 1983), presumably by phosphorylation of membrane proteins. We attempted to determine whether this is caused by an increase in the number of functional Ca channels. Ca channel activity was recorded in myocytes from bovine, cat and guinea-pig hearts. Patch clamp currents, before as well as during adrenaline perfusion or cAMP-injection, were averaged to obtain I_{si}. In patches containing only one channel, I_{si} increased although the single channel current remained unchanged and no new channels appeared. Open and closed time histograms of the current and analysis of the non-stationary current fluctuations suggested that β-adrenergic stimulation enhanced I_{si} by an increased probability of the individual Ca channel being open rather than by an increased number of channels.

DNA Synthesis in cultured ventricular cells of the adult rat. S. L. Jacobson (Department of Biology, Carleton University, Ottawa, Ontario, K1S 5B6, Canada).

Cell cultures of ventricular cardiomyocytes of the adult rat were examined for DNA synthetic activity. The cultures were prepared by a variation of the method of Jacobson (1977). Synthesis was determined by ^{3}H-thymidine incorporation assayed by autoradiography. In the assay, myocytes were distinguished by an immunohistochemical technique that identified cells in culture that contained myosin.

The rate of DNA synthesis in the assayed cultures was markedly lower than that reported for cultures of adult rat ventricular cardiomyocytes prepared by the method of Claycomb and Bradshaw (1983). Some possible reasons for the difference are discussed.

Is there an "oxygen paradox" on the cellular level? P. Schwartz, H. M. Piper, R. Spahr, J. F. Hütter and P. G. Spieckermann (Zentren Anatomie und Physiologie, Kreuzbergring, Humboldtallee 23, D-3400 Göttingen, F.R.G.)

Cultured adult cardiac myocytes (JMCC 14: 397 [1982]) were exposed to anoxia under substrate-free conditions and then reoxygenated. In anoxic cell culture metabolic changes develop similarly to those in arrested hearts. Release of cytosolic enzymes (LDH, MDH) starts with minor energetic disturbances and proceeds closely correlated to the actual ATP content, while loss of mitochondrial (GLDH) and lysosomal enzymes (acid phosphatease) is not seen. Below 2 μmol ATP/g_{ww}, an increasing number of cells become irreversibly damaged; above this level, 30 min reoxygenation leads to an extensive recovery of the whole preparation. The results indicate that early leakage of cytosolic enzymes is due to a gradual protein release from the individual cells and is related to reversible membrane alterations, e.g. protrusion of microblebs and extrusion of subsarcolemmal vesicles. Reoxygenation does not induce changes considered typical of the "oxygen paradox", e.g. massive enzyme release, formation of mitochondrial densities and contraction bands. Since mechanical cell-cell interactions are absent in this model, it is suggested that aggravation of heart tissue damage by late reoxygenation is mainly caused by mechanical forces.

[1]) This work was supported by the Deutsche Forschungsgemeinschaft.

Inward rectification of single potassium channels[1]). G. Trube and J. Hescheler (Max-Planck-Institut für biophysikalische Chemie, D-3400 Göttingen and II. Physiologisches Institut der Universität des Saarlandes, D-6650 Homburg, F.R.G.)

K^+-currents in the membrane of ventricular myocytes of the guinea-pig were recorded by using the patch-clamp technique. With 145 mM KCl at the extracellular side of the membrane patch, inwardly directed single-channel currents were observed at potentials more negative than the equilibrium for K^+ ($E_K = 5$ mV), whereas measurable outward currents did not exist (Sakmann and Trube, 1984, J Physiol 347: 641). Histograms of the channels' closed time durations followed a multi-exponential distribution indicating the existence of several classes of closed states. At potentials close to E_K sojourns in a shut state of short lifetime were more frequent and longer than at more negative potentials (mean lifetime = 3.3 ± 0.8 ms at -15 mV, temperature = $21°$ C). The percentage of time spent in the closed state increased about tenfold between -25 mV and -5 mV. Thus, inward rectification seems to be due to a fast gating mechanism which is steeply voltage dependent between E_K and $E_K - 30$ mV.

[1]) Supported by the DFG, SFB 38 "Membranforschung".

Author Index

Subject Index